A CASEBOOK OF ENVIRONMENTAL ISSUES IN CANADA

Monica E. Mulrennan

Department of Geography, Concordia University

JOHN WILEY & SONS, INC.
New York • Chichester • Weinheim • Brisbane • Singapore • Toronto

ISBN 0-471-19964-8

Printed in the United States of America

10 9 8 7 6 5 4 3 2

Printed and bound by Bradford & Bigelow, Inc.

Preface

The idea for this book originated in my sense of frustration at the dearth of ready-to-hand texts that focus on the teaching of Canadian environmental issues. When I complained about this deficiency to a visiting representative of John Wiley & Sons, I unwittingly precipitated an invitation to do something about it.

This casebook is intended to provide undergraduate students from introductory to intermediate levels with a range of analyses that are representative, topically and regionally, of the Canadian scene. I have attempted to present descriptive and interpretive content clearly, while meeting complex issues head-on, and to avoid the sort of pre-digested accounts that "talk down" to students. The casebook takes an interdisciplinary approach, as demanded by environmental problems. While most obviously appropriate for courses in environmental studies, it will also appeal to instructors in geography, biology, anthropology, political science, Canadian studies, resource sciences, development studies, and no doubt other fields and disciplines. While the book aims in the first instance to meet a pedagogical need in Canada, the ten case analyses grapple with problems, structures and processes that are connected to, and recurrent in, other parts of the world. I hope and believe that there is much of value in these pages for international readerships.

The casebook approach is to illustrate and apply certain concepts by addressing "real world" issues. In working through these issues, students gain an understanding not only of natural processes, but of the relationship of scientific understanding to domains of policy and wider social action. Environmental problems – and solutions – are driven largely by economic, social and cultural forces. Students of environment, whether on the biophysical or the human/social side of the great knowledge divide, need to understand that scientific approaches never operate in a political vacuum. Decisions about environment are shaped as much by economy, politics and culture as by natural structures and processes. And science itself is deeply conditioned by these realities.

My own approach, then, has been to explore the interplay between the scientific findings and their socio-environmental contexts – to choose socially situated objectivity over detached neutrality. While my own positions will be transparent enough to the reader, my primary aim is to raise issues and encourage discussion rather than win students over to my perspective. We all value the development of critical thinking skills in our students, perhaps above any other goal in teaching. Environmental issues are a marvelous ground for teasing hidden premises into the open, for discarding dogma, and for leading students to realize that there are no straightforward recipes.

Ten cases were chosen for this book. They focus on contemporary, generally politically controversial issues in managing environment. Some of the episodes addressed herein are ongoing; others have reached resolutions of sorts, perhaps affording the benefit of some hindsight and distance. I have tried in most instances to combine natural and social science information and insights. I invoke the perspectives of different social groups and divergent knowledge traditions (both scientific and local) to foster a certain creative dissonance. Often, students will be led into critical reflection about research methodologies and the limitations on what we can know, and to confront the fact that we must nevertheless find within our knowledge, in the midst of large uncertainties, the wherewithal to act.

The cases reflect regional diversity – east and west coasts, the arctic and subarctic, the prairies, the western montane, the Great Lakes and St. Lawrence. They treat environmental resources that bear on a range of key industries – agriculture, energy, fisheries, forestry, and tourism. They cover the spectrum from remote wilderness to urban industrial environments. At the same time, a spectrum of social scale is implicated, in tandem with the spatial. The question of indigenous and other forms of local and decentralized management versus central state management (together with the compromise of co-management) is a recurring theme. Several of the cases examine transnational interests, politics and negotiations as they bear on issues "within" Canadian boundaries.

The themes that run through the casebook reflect, I hope, a certain coherence that stems not only from my particular orientation but from circumstances generally intrinsic to environmental and resource management. One such theme, already alluded to, is the difficulty of decision-making in the face of uncertainties in ecological knowledge. Students come to recognize that it is not a simple matter of good guys versus bad guys in environmental research and management. The lack of strong information on many wildlife populations, for example, is not necessarily a symptom of negligence, lack of care, or willful distortion. It may not even be caused by inherent methodological limitations – just as often, it relates to the economic reality that gathering good ecological information can be very expensive.

A further theme is the problem of politics dictating policy in the absence of scientific certainty. Political outcomes may yield quite contradictory practices, depending on the interests at stake. A precautionary principle of risk-averse conservation management was enthusiastically espoused by government policy-makers in the case of polar bear management, but rejected in the case of seals. Yet in neither case was there strong scientific evidence that conservative management was demanded. In the case of Pacific salmon, where there *is* strong evidence that it is demanded, it is frighteningly elusive.

The need to exercise caution and skepticism in the application of science relates to another theme, which is the importance of the "unofficial" but often sophisticated traditions of knowledge held by indigenous and other people about their local ecologies. This goes hand-in-hand with recognizing the importance of local institutions empowered to manage resources. Social and cultural diversity are not incidentally or accidentally related to biodiversity and sustainability – distinctive institutional practices foster distinctive ecosystemic linkages and renewals. The rights of aboriginal people to distinct systems of property and governance, for example, are an important factor in pursuing environmental objectives.

While several of the case studies display my conviction about the importance of local involvement, it is no panacea, and it frequently presents us with paradoxes. At Temagami, the case the Wendaban Stewardship Authority is cited as a particularly successful community-based initiative, but its future is in grave doubt due to lack of government support. In Hamilton-Wentworth, meanwhile, the municipal government has demonstrated its support and commitment to enable local neighborhoods, organizations, and individual citizens to assume control of environmental agendas, yet several attempts to rally the community have garnered little grassroots involvement or interest.

An overarching conclusion of the casebook is the urgent need for collaboration and cooperation linking all social and spatial scales. In a global society, neither the centralized institutions of state governance nor local institutions can go-it-alone in the pursuit of environmental agendas. Significant social transformations are needed to dissolve outmoded jurisdictional prerogatives, if local communities are to play the role that they can and must. Small-scale communities can be guardians of local resources in ways that are impossible for the state to match. And science itself must forego some of its role as the privileged idiom of modern bureaucratic management, to pursue the less authoritative but hugely more constructive approach of learning with people on-the-ground. Deep histories of cultural knowledge and experience *in situ*, together with life adaptations that yield more routine empirical exposure than is frequently feasible on limited scientific budgets, make local people indispensable allies in our quest for understanding.

It is my wish that this book will be a vehicle for students, and all interested readers, to share in the construction of a culture of environmental inquiry and action that advances these goals.

Monica E. Mulrennan
Montréal, July 1997

Acknowledgments

The preparation of this Casebook was made possible by the cooperation and assistance of numerous people from government agencies, environmental organizations, academic and research institutions, and businesses. In addition to providing documents and other research material, the following people shared their expertise, experiences and perspectives on the issues covered. I am particularly indebted to the several individuals who gave interviews and reviewed drafts of the cases:

Rebecca Aldworthy, Concordia Animal Rights Association, Montréal, Québec
Rolfe Antonowitsch, Prairie Farm Rehabilitation Administration, Regina, Saskatchewan
Steven Atkinson, Dept. of Renewable Resources, Wildlife, & Economic Development, Yellowknife, NWT
Philip Awashish, Grand Council of the Crees/Cree Regional Authority, Montréal, Québec
Sid Barber, Sustainable Land Management, Govt. of Saskatchewan, Regina
Jane Barr, Commission for Environmental Cooperation, Montréal, Québec
Mark Bekkering, Regional Environment Dept., Regional Municipality of Hamilton-Wentworth, Ontario
André Bourget, Canadian Wildlife Service, Québec, Québec
Normand Châtelier, Computer Consultant, Montréal, Québec
Dean Cluff, Dept. of Resources, Wildlife and Economic Development, Yellowknife, NWT
Brent Coat, Parks Canada, Ottawa, Ontario
Brian Craik, Grand Council of the Crees of Québec, Ottawa, Ontario
Rick Cuciurean, Cree Trappers' Association, Eastmain, James Bay, Québec
Jim Devries, Institute for Wetland and Waterfowl Research, Ducks Unlimited Canada, Stonewall, Manitoba
Alan Dextrase, Ontario Ministry of Natural Resources, Peterborough, Ontario
Kathy Dickson, Canadian Wildlife Service, Hull, Québec
René Dion, Cree Regional Authority, Montréal, Québec
Charles Drolet, Canadian Wildlife Service, Québec, Québec
Patricia Dwyer, Canadian Wildlife Service, Hull, Québec
Harvey Feit, Department of Anthropology, McMaster University, Hamilton, Ontario
Beth Halford, ConservaPak Seeding Systems, Indian Head, Saskatchewan
Jim Halford, ConservaPak Seeding Systems, Indian Head, Saskatchewan
Sandy Hunter, Ontario Native Affairs Secretariat, Ottawa, Ontario
Julia Innes, Canadian Wildlife Service, Québec, Québec
Laurie Kingston, International Fund for Animal Welfare, Ottawa, Ontario
John Kort, Prairie Farm Rehabilitation Administration, Shelterbelt Centre, Indian Head, Saskatchewan
Larry Lenton, Prairie Farm Rehabilitation Administration, Regina, Saskatchewan
Doug McKell, Saskatchewan Soil Conservation Association Inc., Indian Head, Saskatchewan
Blair McClinton, Saskatchewan Soil Conservation Association Inc., Indian Head, Saskatchewan
Lee Moats, Ducks Unlimited Canada, Regina, Saskatchewan
Jim Morrison, Wendaban Stewardship Authority, Haileybury, Ontario
Ted O'Brien, Prairie Farm Rehabilitation Administration, Regina, Saskatchewan
Alan Penn, Cree Regional Authority, Montréal, Québec
Norman Ragetlie, Regional Environment Dept., Regional Municipality of Hamilton-Wentworth, Ontario
Austin Reed, Canadian Wildlife Services, Québec, Québec
Dick Russell, Canadian Wildlife Service, Hull, Québec
Lorne Scott, Dept. of Environment and Resource Management, Govt. of Saskatchewan, Regina
Terry Scott, Saskatchewan Agriculture and Food, Regina, Saskatchewan
André Savoie, Parks Canada, Ottawa, Ontario
Mitchell Taylor, Dept. of Resources, Wildlife and Economic Development, Yellowknife, NWT
Ted Weins, Prairie Farm Rehabilitation Administration, Regina, Saskatchewan
Steve Wendt, Canadian Wildlife Service, Hull, Québec
George Wenzel, Department of Geography, McGill University, Montréal, Québec
Floyd Wilson, Prairie Farm Rehabilitation Administration, Regina, Saskatchewan

In addition to the institutions represented by many of the above individuals, assistance provided by the following agencies has also been very helpful and much appreciated:

Canadian Arctic Resources Committee
Canadian Heritage, Parks Canada
Canadian Parks and Wilderness Society
Department of Fisheries and Oceans
Department of Indian Affairs and Northern Development
Environment Canada
Ontario Ministry of Natural Resources

My sincere thanks to the faculty, support staff and students of the Department of Geography, Concordia University, Montréal. I am particularly grateful to Dr. Patricia Thornton, Department Chair, for her encouragement and support for this project; and to my colleague, Dr. Geraldine Akman for her comments on earlier drafts of several of the cases. My thanks to Paul Wrigglesworth, Khatoon Abbasi, and Luc de Montigny for cartographic assistance, and to Dr. Jacqueline Anderson for her advice on map design. Thanks also to Annie Pollock and Tina Skalkogiannis for redirecting calls and faxes, and for generally helping to allow me the space necessary to complete this project. To my students I owe a debt of gratitude – their queries and insights over the years have helped shape many of the discussions presented here.

I am grateful to Daniel Gingras, John Wiley & Sons, for getting me started on this project, and to Jennifer Yee, also of John Wiley & Sons, for guiding me through the various stages.

Very special thanks go to Laurie-Anne White, my Research Assistant, who has worked closely with me on this project over the past months, who enabled me to stay home to write while she tracked down volumes of source materials and contacts, and who handled many of the logistics of assembling the final manuscript. Her resourcefulness and consideration have been exceptional.

My deepest appreciation goes to my husband, Colin Scott, Department of Anthropology, McGill University, who discussed with me key issues in regard to each case and provided valuable editorial advice. With a looming publication deadline, his assistance included co-authorship of the Canada Goose case study.

The final word goes to my three month old son – Sean Francis Scott – who arrived in the middle of this project and has yet to experience "normal" family life.

About the Author

Monica E. Mulrennan is an Assistant Professor in the Department of Geography at Concordia University, Montréal, where she teaches undergraduate and graduate-level courses on environmental issues, indigenous resource management, and landscape evolution. She was born and educated in Ireland and in 1990 completed a doctoral degree in Geography, specializing in coastal geomorphology, at University College Dublin.

Shortly afterward, she moved to Australia where she held postdoctoral positions at the University of Wollongong, New South Wales and at the Australian National University's North Australia Research Unit (NARU) in Darwin. During this period, she extended her research to the evolution and development of tropical estuarine environments. While in northern Australia, she developed an interest in indigenous knowledge, aboriginal rights in environmental policy, and co-management of coastal environments. She has worked particularly closely with indigenous Islanders and organizations in Torres Strait, northern Queensland.

Dr. Mulrennan moved to Canada in 1993 on a Natural Sciences and Engineering Research Council international postdoctoral fellowship at McGill University, and a year later took up an assistant professorship at Concordia University. She maintains her research involvement at Torres Strait and is pursuing comparative research into Cree knowledge of coastal environments, indigenous land and sea tenure systems, and environmental change at James Bay, northern Québec.

Her monographs include *A Marine Strategy for Torres Strait: Policy Directions* (NARU, Australian National University, 1994) and *The Geomorphology of the Lower Mary River Plains, Northern Territory* (NARU/Conservation Commission of the Northern Territory, 1993). She has published papers in the *Journal of Coastal Research*, the *Journal of Sedimentary Geology*, and various edited volumes.

Table of Contents

List of Figures

List of Tables

Case One
Fraser River: The Mystery of the Missing Sockeye

Focus Concept

A combination of environmental stresses, illegal fishing, imperfect science, and inadequate technology creates a strong case for adherence to conservative fisheries management systems.

Introduction

In 1994, 1.3 million sockeye salmon "went missing" from British Columbia's strongest salmon runs, in the Fraser River watershed. Six months later, a public review board headed by former Tory Fisheries Minister and Canada's current Ambassador for the Environment, John Fraser, released the results of its investigation into the mystery. "The message is simple," Fraser declared:

> If something like the 1994 situation happens again the door to disaster will be wide open. One more 12-hour opening could have virtually eliminated the Late run of sockeye in the Adams River [a tributary of the Fraser River]... Unless all parties work together and manage much more competently, the tragedy that befell the Atlantic cod fishery will repeat itself on the west coast.[1]

On the face of it, the cause seemed easy to define - "One more 12-hour opening," perhaps suggesting that the finger of blame should be pointed at the commercial fishing fleet. In fact, Fraser's review board largely exonerated the commercial fleet and instead highlighted the role of high water temperatures and mismanagement of the aboriginal and commercial fisheries.[2] More specifically, Fraser points to the policies of the previous Tory government: "Cutbacks and budget reductions were made to the extent that the department was left in charge without clear lines of accountability or necessary tools to enforce its regulations with any credibility."[3] The department was reduced, he concluded, to "a state of denial as to the existence of a problem ... In this chaos, blame was found everywhere and little attention was paid to the core problem: the system had become dysfunctional."[4]

The missing fish represent the difference between the number of fish passing Mission (a monitoring station, just upstream of the commercial fishery), and the spawning counts plus the Native catch estimates made further upstream. But as fisheries biologist Carl Walters explains, whether the Mission estimate (determined by an out-dated echo-sounding program) was too high, or the Native catch was underestimated, or the fish simply died from thermal stress will never be known. The basis reason is "because investment in monitoring the stocks, B.C.'s largest (!), was hopelessly inadequate."[5] The 1994 Fraser River runs were to end up paying the consequences of this breakdown of monitoring programs. According to Fraser, the result has been the creation of a situation where the sockeye are "now, more than ever before, critically endangered."[6]

---

[1] Fraser River Sockeye Public Review Board, *Fraser River Sockeye 1994: Problems and Discrepancies* (Ottawa: Public Works and Services, 1995) xii.
[2] G. Meggs, *Salmon: the Decline of the British Columbia Fishery* (Vancouver: Douglas and McIntyre, 1991)
[3] Fraser River Sockeye Public Review Board, *op. cit.*
[4] *Op. cit.*
[5] C. Walters, *Fish on the Line: The Future of Pacific Fisheries* (Vancouver: The David Suzuki Foundation, 1995), 26.
[6] Fraser River Sockeye Public Review Board, *op. cit.*, 12.

Yet, recent reports and findings indicate that the resource is already well on its way to recovery. The 1996 run of sockeye to the Fraser River returned in record strength to contribute nearly 1.8 million salmon to the Canadian fishery catch.[7] Some scientists suggest that the potential exists to double existing runs, some of which are as strong as they have been since the earliest days of the commercial fishery.

> Even producing at half its historical level it is the greatest salmon river on the planet. And its potential is amazing. The run of more than 22 million salmon that came back in 1993 could be doubled – and that figure could possibly be doubled again.[8]

Rather than taking false comfort in these assessments, we need to build a clear understanding of some of the factors responsible for what happened back in 1994. How did millions of sockeye salmon simply disappear? What or who was to blame? What has been done since then to assist in the recovery of the runs?

Life Cycle of the Sockeye Salmon

Of the various species of Pacific salmon, the sockeye (*Oncorhynchus nerka*) are the most commercially valuable.[9] The Fraser River is the largest producer of sockeye in the world, supporting approximately 100 stocks. Sockeye salmon returning to the Fraser are the economic backbone of commercial fisheries in both Canada and the United States.

The sockeye are also a source of fascination because of their extraordinary life cycle and finely tuned navigation systems that allow them to make their way thousands of kilometers from the North Pacific back to the precise streams of their origin in the Fraser River watershed. Their extraordinary sense of smell is reputed to aid them in their epic struggle to return to the streams where they were born.

Sockeye generally mature at four years of age, resulting in four distinct year classes or cycles. In the fall, each returning female sockeye digs her nest and deposits about 3,500 eggs in clusters, called redds, in the gravel of the streams where they themselves had been spawned. After immediate fertilization of the eggs by male sockeye, the female spawners, close to death, cover the eggs and guard them for as long as possible. After a six-to seven-month incubation period, millions of baby sockeye, called alevins, emerge from the gravel. About 10 percent survive to the fry stage. They feed and grow in nearby lakes for a year. About 25 percent of these fry survive to become seaward-migrating smolts. On average, about 10 percent of the smolts survive to be adults. This represents about nine fish for 3,500 original eggs.[10]

Their journey to the open ocean is not an easy one. The young sockeye are swept up by the Alaskan coastal current which carries them northward up the B.C. coast and the Alaskan panhandle. From there they swim close to the northern edge of the Gulf of Alaska before sweeping past Kodiak Island and out to the Aleutian Islands. Since sockeye spend most of their life in the open ocean, where they are out of sight and largely unstudied, we know relatively little about this part of their life cycle. It is clear, however, that during this stage they are subject to a combination of factors including unfavorable ocean temperatures and scarcity of foods such as zooplankton, and many do not survive. In addition many succumb to

---

[7] Department of Fisheries and Oceans, 1996 Fraser River salmon season exceeds expectations, News Release, NR-PR-96-42E, 26 September 1996.
[8] M. Hume, *Adam's River: The Mystery of the Adams River Sockeye (*Vancouver: New Star Books, 1994).
[9] B.C. landings of sockeye salmon in 1994 were 30,830 metric tonnes in live weight and valued at $196,300. Total B.C. salmon landings were 65,829 metric tonnes and valued at $257,485 (L.C. Gagnon, Statistical Services, Ottawa: DFO, n.d.).
[10] C. Groot, and L. Margolis, eds., *Pacific Salmon Life Histories* (Vancouver: UBC Press, 1991).

predators – from natural ocean enemies, such as mackerel, to human harvesters. Those that survive (only about one quarter of one percent of the eggs laid by female sockeye at the beginning of the four-year cycle) cover vast distances during two continuous years of ocean travel.

On their return, salmon have two possible routes back to the Fraser River: skirting the west coast of Vancouver Island, into Juan de Fuca Strait and eventually into Georgia Strait and the Fraser River, or along a coastal route on the east side of Vancouver Island through Johnstone Strait (Figure 1.1). Both routes require that they make it past the Alaskan net fisheries, a Canadian troll fishery, and avoid the elaborate obstacle course of hooks, lines and nets along their way. Upon arriving at the Fraser, most of the sockeye move directly upstream, though some – like the Adams River run - may pause for three to six weeks off the mouth of the river before moving upstream for the last leg of their journey.

The sockeye undergo marked physical changes during their upstream migration. The silvery hue of the sea-going sockeye transforms into shades of green and red, and the male develops a marked hump on its back. They stop feeding while in the freshwater environment, and depend on their energy reserves for both locomotion and the development of eggs and sperm. They must overcome several obstacles before they reach their ancestral spawning grounds. In particular, the sockeye must battle the treacherous currents of Fraser canyon before they scramble up fish ladders at Hells Gate, an extremely narrow gorge in the Fraser Canyon, through which most of the Fraser River watershed passes. This is one of the most difficult migration points for returning salmon.

The salmon returning to spawning beds on the tributaries of the Fraser River are categorized into four runs[11]: Early Stuart, Early Summer, Summer, and Late run, which includes the Adams River run. For every 10 mature sockeye returning to spawn, about eight are taken by the commercial, Aboriginal and recreational fisheries, leaving but a single pair to escape to the spawning grounds. The total number of mature salmon that pass through (or escape) the fisheries and return to their rivers of origin to spawn is referred to as "escapement." On average, 10 million sockeye are taken from the Fraser stocks each year; about 18 percent of the catch is taken by United States fisheries.[12] Target escapement levels are set by fishery management authorities for individual stocks of salmon and should ensure maximum sustainable yield (MSY) of the stock.

History of the Fraser River Salmon Fishery

The commercial fishery in British Columbia dates back at least to the period of the fur trade, when First Nations people bartered fish products with the Hudson Bay Company in the 18th and 19th centuries. The establishment of an industrial fishery on the Fraser River in the 1800s brought the First Nations fisheries under the effective jurisdiction of the Canadian state. Cured salmon and other fish were exported, and before long the salmon industry was of greater importance than the fur trade. After the Gold Rush of the 1850s and 1880s the industry shifted from export of salted fish to large scale canning of Pacific salmon. The first cannery opened on the Fraser River in 1866, at New Westminster. The number of canneries on the river increased steadily thereafter, reaching a peak of 49 in 1901.

By the 1890s, the American catch of Canadian-spawned sockeye had become a concern to the B.C. fishing industry. This had more to do with Canadian claims of ownership and conflicting regulatory schemes between the United States and Canada than concern for the conservation of the stocks. Even at the turn of the century, the supply seemed inexhaustible. With such abundance far exceeding industry's capacity, the emerging industry focused on technical and marketing issues and gave little thought to conservation or environmental conditions:

[11] A run refers to a stock of the same species that returns to a river over a particular time period.

[12] Fraser River Sockeye Public Review Board, *op. cit.*

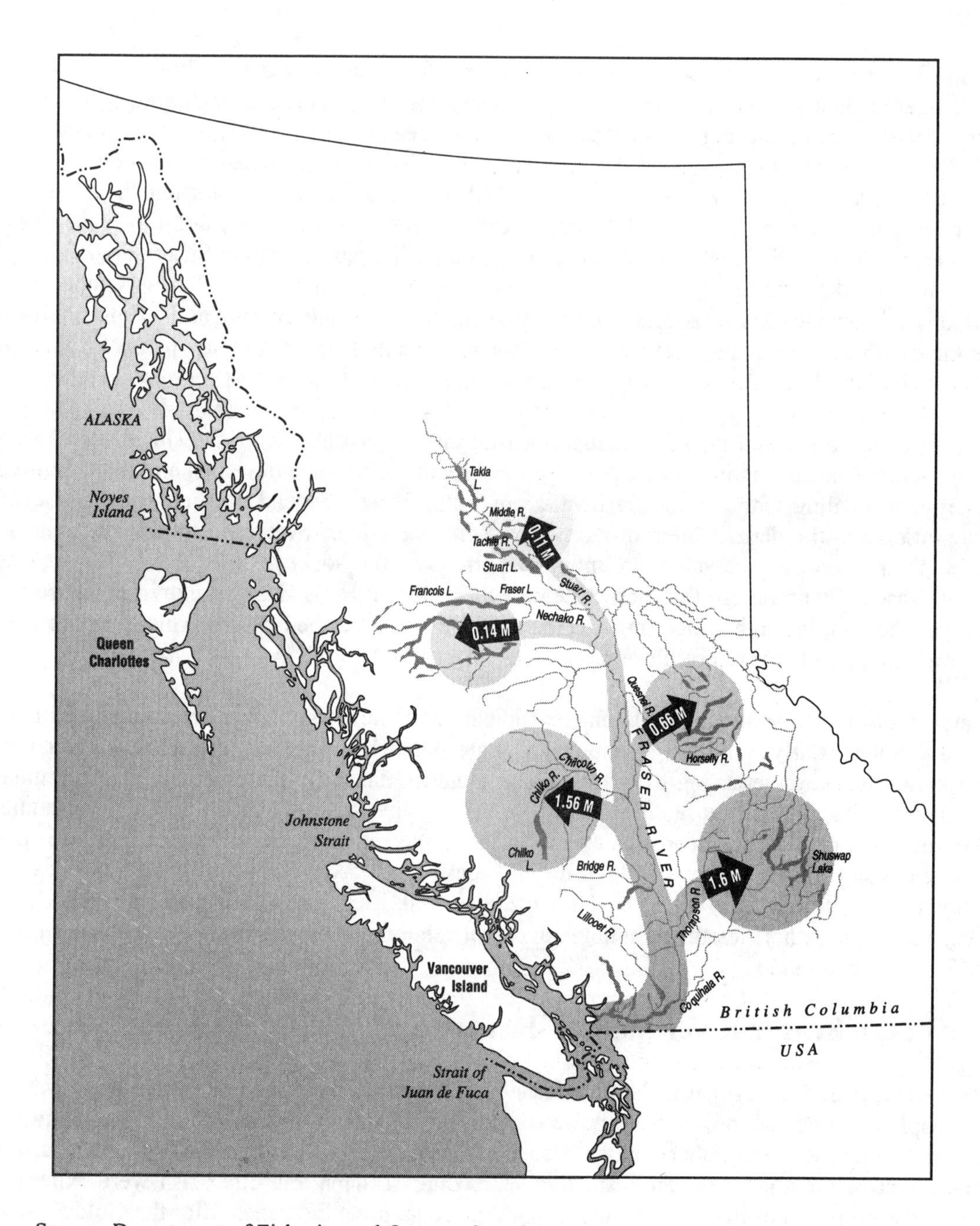

Source: Department of Fisheries and Oceans, Canada

Figure 1.1. Up-River Sockeye Migration in 1994, including Escapement to the Major Spawning Areas.

> ...The catch that year [1901] was so great that every one of the canneries on both sides of the international line filled every can they had or could obtain; and in addition ... millions [of fish] ... which could not be used ... were thrown back dead into the water.[13]

By the next decade it became apparent that the supply of sockeye was limited, and that the long-term viability of the fishery was threatened by increased fishing intensity in both the United States and Canada.

Events at Hells Gate in the Fraser Canyon in 1913 and 1914 proved catastrophic for the Fraser River salmon. Construction by the Canadian Northern Railway along the east bank of the canyon near Hells Gate resulted in large amounts of rock being dumped into the river in 1911 and 1912, changing its flow pattern. In 1913, in spite of a record commercial catch of over 32 million fish, observers noted huge numbers of sockeye jammed below Hells Gate, unable to reach their spawning grounds. This set the stage for a staggering decline in future production. To address the problem, efforts began in 1913 to remove some of the rocks from the canyon, but a huge rockslide along the east bank dumped an additional 100,000 cubic yards of rock into Hells Gate in 1914. With their passage blocked millions of salmon died. By March 1915, more than 60,000 cubic yards of rock were removed and it was believed that river conditions had been restored. No further man-made changes to Hells Gate were made until the 1940s after biologists documented their findings that the average commercial catch on the Fraser River sockeye run was now only 25 percent of that prior to the 1913-14 catastrophe.

A long rebuilding process over the past decades has increased the Fraser River sockeye runs, but overall they are still below historic averages. Multiple environmental stresses, ever increasing efficiency within the commercial fleet and unresolved domestic and international policy considerations, pose a challenge to today's efforts to conserve the stocks.

Environmental Factors

Sockeye experience multiple stresses throughout the different stages of their life cycle, because of their dependence on a number of habitats. They start their life as eggs in the gravel of riverbeds; they spend their first year of life in lakes, use rivers to move from one habitat to another, and grow to maturity in the ocean. The quality of each of these habitats is vital to their survival. Three physical features can be identified as critical: open-ocean conditions, river water temperatures and flow rates, and water quality.

Ocean conditions affect the success of the sockeye's life at sea; the availability of food affects their growth rate, while water temperatures and currents affect their migration routes. Inter-annual variability in these conditions will determine their net survival at sea, weight at maturity and details of run timing and diversion rates.[14] Under warm ocean conditions the sockeye tend to avoid the warm offshore waters in favor of the cooler coastal waters of Johnstone Strait. Whether warmer coastal ocean temperatures are related to a global warming phenomenon or to local variability is of interest to long-term management.

River water temperature is the most critical and controversial environmental factor associated with sockeye migration. Adult sockeye migrating through freshwater swim most efficiently at temperatures close to their own body temperature of 15°C. If the surrounding water is colder, their metabolism may slow down. If much hotter, physiological processes speed up, and valuable energy may be wasted. Water temperatures above 21.5°C are usually lethal. River conditions are thus crucial for successful migration – fish must have sufficient energy to reach their spawning grounds. Those encountering both fast waters and temperatures at the upper end of their tolerance are particularly vulnerable, especially if they are

[13] J.P. Babcock, Annual Report of the [B.C.] Commissioner of Fisheries, 1909.

[14] Diversion refers to the act of the fish moving away from a more usual course or migration route.

returning to more distant spawning areas, such as the Early Stuart run more than 1,000 km upstream from the mouth of the Fraser.

Additional stresses will arise for sockeye swimming through contaminated water. Water quality in rearing lakes and river migration routes is particularly important at the fry stage and during spawning migration. High-level pollution from industrial and municipal outfalls, such as the Greater Vancouver Regional District's Annacis Island outfall, has been identified as potentially harmful to migrating salmon. The precise effects of such stresses are unknown but it is recognized that they can delay the sockeye's migration, deplete their energy reserves and render them more vulnerable to subsequent stresses, such as encounters with nets.

At a more general level there is the issue of policy regarding environmental protection of the Fraser River basin; this has direct implications for the sustainability of the salmon resource. Over the years various physical impediments have been removed and high currents reduced by the construction of suitable fish ladders. However, a dam on the Nechako River continues to disrupt the natural regime of the river. Environmental policies need to be broadened to include the regulation and control of water temperatures; this would involve careful land use and forestry practices to prevent rapid warming of precipitation flowing to the river basin as well as proactive cooling of the more critical rivers. Meggs documents several practical environmental initiatives, which have been taken in recent years.[15] These, he suggests, reflect a new level of public commitment to save the salmon and include the marsh and wetland clean-up projects undertaken on the Fraser by unemployed fishermen, organized by the T. "Buck" Suzuki Environmental Foundation. More ambitious initiatives include: a local pollution campaign which successfully eliminated two toxic organochlorines from pulp effluent; a province-wide movement that forced the cancellation of Alcan's Kemano Completion Project after hundreds of millions of dollars had already been spent; and an international effort to clear the North Pacific of the driftnet fisheries that were targeting salmon. It was salmon too which provided the impetus for the recent establishment of a new *Forest Practices Act*, a world standard in environmentally sustainable forest practices.

Institutional Structure of the B.C. Salmon Fisheries

The federal government has jurisdiction over "sea coast and inland fisheries" under section 91(12) of the *Constitution Act*, 1867. This jurisdiction is exercised by the Department of Fisheries and Oceans (DFO), which manages and controls fisheries through the provisions of the *Fisheries Act* and associated regulations. The primary objective of DFO is to conserve the salmon resource by ensuring that enough fish of each species reach the spawning grounds to ensure that the population can at least be maintained, if not enhanced. Fish not needed for spawning are then allocated to various competing user groups. The aboriginal food fisheries have priority by virtue of section 35(1) of the *Constitution Act*, 1982. Commercial and recreational fisheries follow next in priority.

Day-to-day management of the Fraser River salmon stocks is the responsibility of the DFO except in areas under the jurisdiction of the Pacific Salmon Commission (PSC), established under the *Pacific Salmon Treaty* 1985 between Canada and the United States. DFO develops a fish management plan before each year's fishing season, based on crude preseason forecasts of abundance each year (+/-50%). Final approval of all fishing plans rests with the Minister of Fisheries and Oceans, who has wide discretionary powers under the *Fisheries Act* and other legislation to restrict fishing times and places, issue and withdraw licenses, and take legal steps against individuals or groups who pollute or degrade fish habitat.

---

[15] G. Meggs, *op. cit.*

Case One

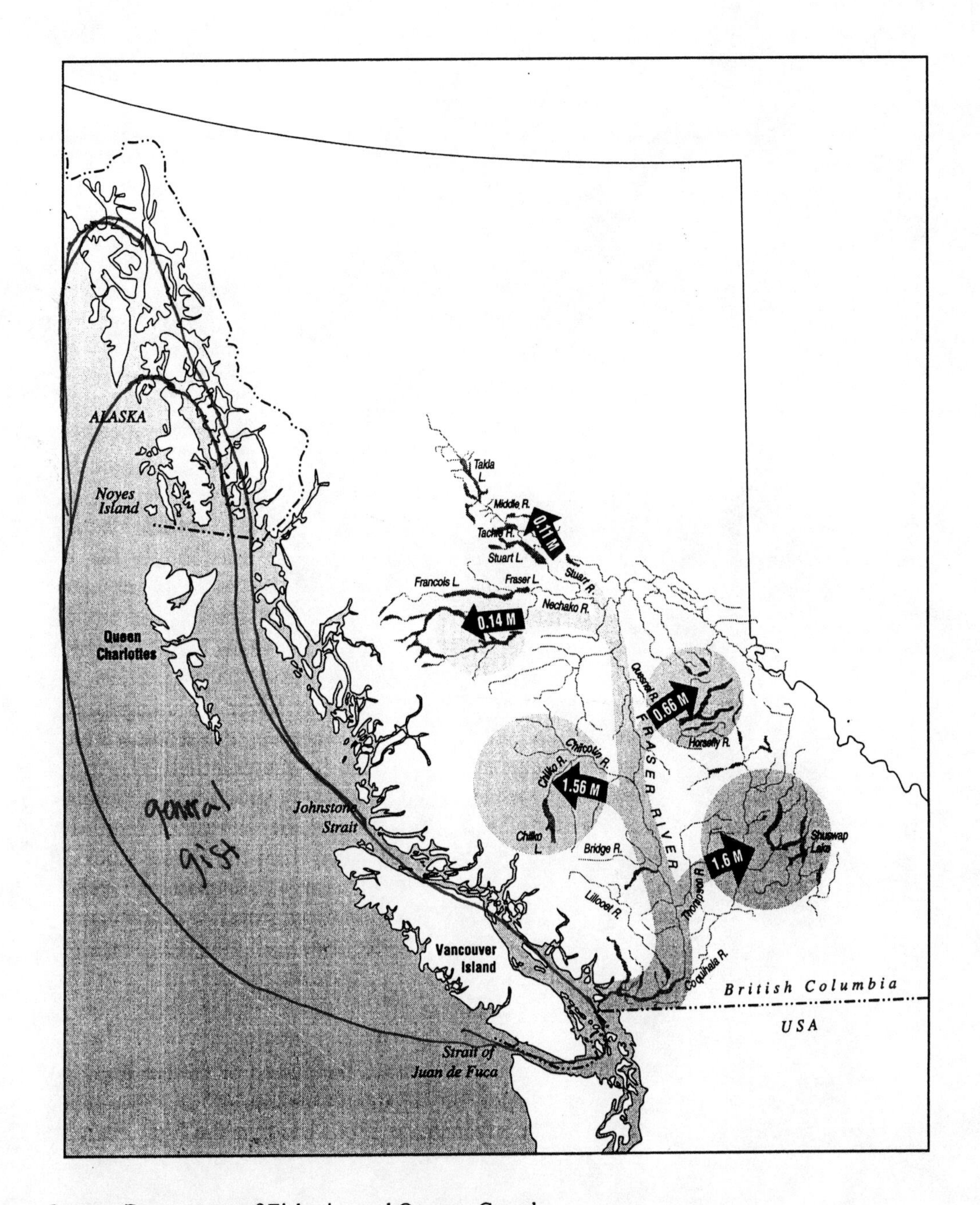

Source: Department of Fisheries and Oceans, Canada

Figure 1.1. Up-River Sockeye Migration in 1994, including Escapement to the Major Spawning Areas.

Since fisheries and navigable waterways are a federal responsibility, while lands and other land-based resource industries are under provincial jurisdiction (under section 92), the provincial government has an ambiguous and easily avoided role in fisheries. While provincial legislation relating to forestry, mining and other land uses affects marine and freshwater habitat, the impacts are often left to the federal government to resolve. In turn, the federal government is constrained by provincial priorities and legislation. It can make rules and regulations on the capture of fish but is unable to control the context of fishing. As a result, fisheries management has been fraught with inter-jurisdictional squabbles and conflicts, which have involved powerful voices from forestry and mining as well as provincial and federal governments. The Canada-British Columbia Agreement on the Management of Pacific Salmon Fishery Issues, signed in April 1997, will provide B.C. with a greater voice in fisheries management. While the agreement does not provide any formal transfer of authority from the federal to the provincial government, "British Columbia will gain a negotiating forum and a more hands-on role for conservation and enforcement."[16] The federal government will retain final say over sensitive issues such as the size of the catch and licensing.

The interests of the various types of commercial fishing (seine, gill net, troll, trawlers, etc.) are represented by several organizations that have been established in recent years, while sport fishers are represented through the Sport Fishery Advisory Board.[17] These organizations have been given a bigger role in fisheries planning and management in recent years through advisory or co-management committees and boards that meet with DFO staff. Proposed changes to federal fisheries legislation promise to give these fishing interests an even greater voice in fisheries management through the establishment of mutually agreeable partnerships with the federal government.

Native fishing interests have been strengthened in recent years by the affirmation of aboriginal fishing rights in the 1990 Supreme Court of Canada's decision in the Sparrow Case. The judgement requires that priority be given to aboriginal fishing rights and that Indians be consulted about fisheries regulations and policies. Various associations have since formed to represent local and regional Native interests. DFO helps to co-ordinate their activities and funding through the Aboriginal Fisheries Strategy (AFS). Efforts to bring together all B.C. Indian communities in a fisheries framework agreement or to develop a co-ordinated fishing plan for the Indian groups on the Fraser have so far failed. More recently, Ottawa has attempted to negotiate a fisheries policy with individual bands and tribal councils. The process is difficult, contentious and challenging from a fisheries management perspective, requiring the federal government to find a way to exercise its constitutional obligations to up-river aboriginal groups and to ensure conservation. Fortunately, a growing body of evidence demonstrates the capacity of co-management arrangements between governments and First Nations peoples to address some of the major biological, economic and political problems of renewable resource management.[18]

The migration of various salmon stocks across the international border imposes further complexities and limitations on the ability of the federal government to manage and conserve the fishery. Canada and the United States entered into the *Pacific Salmon Treaty* in 1985 in an attempt to streamline some of the most difficult of these management issues. The PSC was set up as the main implementing body for the Treaty. It provides regulatory and technical advice to the two signatory countries for all stocks which originate in one country but are susceptible to interception by the other. This includes management of harvests of

[16] H. Winsor and R. Howard, B.C. lands bigger role in fishery, *Globe and Mail*, April 17 1997.

[17] C. Walters, *op. cit.*

[18] E. Pinkerton, Local Fisheries Co-management: a review of international experiences and their implications for salmon management in British Columbia, *Canadian Journal of Aquatic Science* 51:2363-2378, 1994; *Cooperative Management of Local Fisheries: New Directions for Improved Management and Community Development* (Vancouver: University of British Columbia Press, 1989).

sockeye salmon destined for the Fraser River and passing through the Juan de Fuca Strait. The PSC is not supposed to concern itself with domestic allocations within either nation.

The Treaty rests on two principles, conservation and equity; the conservation principle ensures that there is an ample resource, while the equity principle establishes a framework for allocating salmon catches – an essential element for conservation. While the Treaty has provided a much needed forum for consultation and negotiation, a long standing dispute between the two signatory parties over the implementation of the equity principle has reduced the production potential of both countries and has resulted in serious conservation problems.

According to Walters, the establishment of a variety of co-management committees, together with development of the PSC allocation system, and increasing concern for Native allocation of salmon have resulted in "the development of an increasingly complex and rigid set of targets or goals for allocation of catches."[19] This situation has been most pronounced on the Fraser River where aggressive pressure for higher catch allocations has resulted in extremely tight and potentially risky percentage allocations. While recognizing the risks involved in attempting to meet these commitments, particularly in the face of "uncertainty about and variability in the availability of fish,"[20] DFO and PSC have tried to accommodate the various interests by turning to technical tools to improve forecasts of the abundance, distribution and timing of the fish. The preliminary preseason forecast of abundance, from which DFO develops its annual fish management plan, is revised during the fishing season to ensure the highest possible allocation while also achieving escapement goals. A variety of field abundance indices (catches, catch rates, etc.) are used in combination with past experience on the spatial distribution and timing of the migration to revise these estimates.[21] This approach has been relatively successful over the years, largely "because abundances have often been higher than expected and the fish have arrived on time or later than expected."[22] A variety of factors and events combined in 1994, which were beyond the capacity of this management system. The reasons for this failure and the consequences of the breakdown are discussed below.

1994 Sockeye Run

The spawning escapement to many of the major Fraser River tributaries had been good in 1990, and the expectation for 1994 (the fourth and final year of the cycle) was generally optimistic with a provisional estimate of about 30 million fish. This figure was later adjusted downward to 19 million fish after examination of freshwater survival indices, climatic factors and other variables related to run strength. Even so, the 1994 sockeye run was expected to be one of the largest in recent history.

During 1994, warmer than average spring and early summer ocean surface temperatures, attributable to the El Nino phenomenon of 1992 and 1993, caused ocean sockeye to move further north into Alaskan waters. Government and scientists speculated that these conditions would result in the diversion of about 68 percent of the 1994 run into Johnstone Strait.

Management in 1994 was also complicated by the lack of an agreement between Canada and the United States on the allocation of the Fraser River salmon resource under the Pacific Salmon Treaty. In the absence of a united international plan for the harvesting of the stocks, the Minister of Fisheries and Oceans announced an alternative fishing plan, known as Canada's "aggressive fishing strategy," on July 28, 1994. The plan represented an overt strategy to minimize the U.S. share of the Fraser River catch through the positioning of the Canadian commercial fleet on the west coast of Vancouver Island where it

[19] C. Walters, *op. cit.*, 19.
[20] *Op. cit.*
[21] *Op. cit.* 16.
[22] *Op. cit.* 20.

was ready to intercept, ahead of United States fishers, the sockeye predicted to move down into the Juan de Fuca Strait. Despite the expected large diversion of sockeye down the east coast of the island, through Johnstone Strait, it was thought that these stocks were less vulnerable to United States interception and thus could be used to meet in-river catch and spawning escapement requirements.

Early signals quickly confirmed that a substantial component of the Early Stuart and Early Summer runs would indeed migrate down Johnstone Strait. They also indicated that these runs were not as strong as anticipated; the early Stuart run estimate was downgraded in mid-July from 400,000 to 200,000, and the Early Summer run estimate was reduced from 1.1 million to 800,000. By the second week of August, DFO scientists determined that preliminary spawning estimates for the Early Stuart run were in fact closer to 30,000.

Meanwhile, at the beginning of July, as the Early Stuart and Early Summer runs were migrating up the Fraser River, fishing agreements for the 1994 season had yet to be concluded with the Lower Fraser First Nations communities. At the same time, water temperatures near Hells Gate rose to 2°C above the long-term average of 15.7°C and on some days approached the lethal range for sockeye (above 21.5°C) in the Nechako and Stuart rivers. DFO technical reports suggest an en-route mortality of half a million fish as a result of these temperatures.[23]

As for the Summer and Late runs, data available by late August led to a downgrade of the Summer run size from 10.3 million to 6.8 million, while the Late run was upgraded in early September from 7.1 million to 9.3 million sockeye, 3 million of which were believed to be off the mouth of the Fraser River in Georgia Strait. As late as the first couple of weeks of September, estimates for the numbers of fish passing the Mission hydroacoustic station for the Early Stuart, Early Summer and Summer runs were thought to be in line with adjusted in-season targets. By mid-September, however, DFO announced that spawning escapement estimates plus Aboriginal catch estimates for these runs were 1.3 million lower than the number of sockeye anticipated, based on PSC estimates at Mission for these runs.

As the commercial fishing season progressed, catch and effort data indicated an ever-increasing diversion of sockeye (up to 90 percent) down Johnstone Strait. This required Canadian fishery managers to abandon plans to make major interceptions in the southern areas and to reposition themselves further to the north and in Johnstone Strait. Seine fishing in Johnstone Strait took place over four short periods between August 8 and 31. Even in the face of an in-season alteration of fishing strategies, the mobility and fishing power of the commercial fleets were effective in harvesting about 78 percent of the sockeye bound for the Fraser River. United States fishers took about one sixth of this harvest. Both the United States and Canadian commercial fisheries were closed by early September, having met their allocations. The 1994 harvest, including the Fraser River First Nations fishery, amounted to about 80 percent of the total run.

The crisis was deepened in late September when the PSC announced a revision of its estimate of the Adams River and other Late run sockeye. The revised figures suggested these runs were unlikely to exceed 1.5 million fish, half of the in-season estimate. Final estimates for Late run stocks, released in January 1995, showed the Adams River escapements set back to pre-1940 levels.

By the end of the season, claims and counter claims were being made, blaming illegal fishing, bad management, out-dated technology and environmental disaster for the "missing fish" and poor escapement. The credibility of the various management agencies was seriously undermined. In an effort to regain some control over the situation, the Minister announced the creation of an independent review board – the Fraser River Sockeye Public Review Board, headed by John Fraser, to investigate the salmon

[23] DFO Technical Working Group, 1995.

shortfall. The terms of reference of the review board were later extended to include an examination of the Pacific salmon more generally.

Problems and Discrepancies on the Fraser River

Three years after the events of 1994 no one, neither the authorities, the experts nor the Sockeye Public Review Board, knows precisely what happened on the Fraser River or exactly how it happened. The Board, after conducting a thorough appraisal of the methodology, estimates and potential errors, felt that the information available did not warrant a further exercise in "accounting for the unaccounted."[24] Nevertheless there is general agreement that action must be taken to ensure that what happened in 1994 does not happen again.

The findings of the Fraser River Sockeye Public Review Board also suggest that while illegal fishing and environmental stresses, such as the unprecedented diversion of returning stocks and unfavorable in-river conditions in 1994, complicated the task of fisheries managers, "an over-reliance on the quality of historic in-season estimates, and an optimistic attitude regarding run size, fostered risky management decisions."[25] Furthermore, downsizing in the federal government resulted in the restructuring of DFO and dramatically affected the regulatory and enforcement situation in the fisheries. This in turn lead to an increased unreliability in catch estimates and a particularly large unreported catch.

While the Board was unable to pinpoint the cause of the 1994 crisis, it concluded that the following factors aggravated the resource management problems of 1994:

1. A breakdown in enforcement or the perception of a breakdown in enforcement;
2. Failure to secure timely and comprehensive arrangements with First Nations peoples;
3. Anomalous ocean and in-river environmental conditions;
4. Significant errors in estimating in-season marine run and stock sizes leading to an inability to deliver adequate fish to meet spawning escapement targets and in-river catch allocations[26];
5. Delays in accurate catch reporting;
6. Flawed management structures and communications between DFO, PSC and First Nations groups as well as within DFO;
7. Deteriorating staff morale in enforcement and other work areas; and
8. An indeterminate level of illegal marine and freshwater fishing.

Walters echoes many of the findings of the Board[27]. He is particularly concerned about the loss of field monitoring and research programs, which provided essential trend information to harvest management planning and contributed to the refining of forecasting methods:

> Loss of basic abundance information leaves us with little but arm-waving about the importance of such values, and will be catastrophic in the future as growing pressure on watershed resources forces harder tradeoffs and more hard-nosed accounting of economic impacts of alternative policy choices.[28]

[24] Fraser River Sockeye Public Review Board, *op. cit.*, 39.
[25] *Op. cit.*
[26] Different techniques (visual count, fence enumeration, mark-recapture) have inherent positive and negative biases. For example, the Early Stuart run comprised small stocks which were enumerated by visual surveys which could be subject to biases of more than 50 percent, accounting for 170,000 of the "missing fish".
[27] C. Walters, *op. cit.*
[28] *Op. cit.*, 26

There is little question that management of the Fraser River salmon was also complicated by the inability of the two signatories to the *Pacific Salmon Treaty* to reach an agreement for the 1994 season. The exchange of technical information continued, but Canada's decision to pursue an "aggressive fishing policy" seemed highly irresponsible in retrospect. Although the high diversion rate of sockeye through the Johnstone Strait resulted in the ultimate failure of this strategy, it "did, however, contribute to a grab all attitude in the Canadian commercial fleet, and a corresponding removal of any moral responsibility for conservation on the United States side. These difficult circumstances helped create the 12 hours from disaster scenario."[29]

The Board made 35 recommendations for improving the system, including the following:

- That conservation be the primary objective of fisheries managers and all others participating in the fishery;
- That an independent Pacific Fisheries Conservation Council be established to act as a public watchdog for the fishery;
- That DFO and PSC adopt a risk-aversion management strategy, which would factor in the uncertainties inherent to stock estimates, in-season catch estimates and environmental problems;
- That clearer lines of authority and accountability be established between DFO in Ottawa and the Pacific Region;
- That better communications between the various government organizations and First Nations be developed, including the involvement of First Nations communities in the in-river management of the fish;
- That better communications be established between commercial and recreational fishing sectors;
- That DFO recognize enforcement as an essential element of the fishery management process and institute a plan to ensure that an effective level of enforcement is re-established; and
- That DFO develop effective environmental and development plans and policies to protect the Fraser River basin.

Recent Initiatives

On March 29, 1996 the federal government announced the establishment of the Pacific Salmon Revitalization Strategy (commonly known as the Mifflin Plan after current DFO Minister Fred Mifflin), with the primary objective of reducing the size of the salmon fishing fleet. This was considered necessary for four reasons: 1) a change in ocean conditions, which led to lower salmon productivity; 2) short crowded openings, creating increased resource risks; 3) rising costs for fishing fleet operations; and 4) lower world prices for salmon. The main elements of the strategy include:

- A cautious approach to setting harvest levels;
- A 50 percent reduction in the commercial salmon fleet over several years;
- Area licensing,[30] single gear licensing,[31] and license stacking[32] to promote fleet reduction and enhance manageability;

[29] Fraser River Sockeye Public Review Board, *op. cit.*

[30] To reduce the number of vessels in any given fishery, fishers are now required to choose a specific area for a four-year period. For seiners, there are two areas – one in the north and one in the south – while there are three areas for gillnet and troll fishers.

[31] To promote long-term viability, licence holders are required to choose a single type of gear that they would continue to use on a permanent basis.

- Adjustment assistance for displaced fishers;
- New consultative mechanisms; and
- Development of a transparent and open allocation policy between sectors and within the commercial sector.

According to DFO, the success of the strategy to date is reflected in the fact that there were 30 percent fewer boats on the water at the start of the 1996 season. Conservation measures, including 'risk averse' management plans, no in-river fishing on the Early Stuart run, and an agreement between Canada and the United States to share in the management and harvest of the 1996 stock, are credited with allowing DFO to achieve or surpass spawning escapement targets for most salmon stocks.[33] For example, the 1996 escapement on the Early Stuart was the second highest recorded for this cycle and exceeded the spawning escapement target by 33 percent. The smaller fishing fleet was able to harvest the available catch at lower cost, resulting in an increase of $21 million in net income despite lower salmon prices.

The Minister of Fisheries and Oceans announced additional elements of the strategy in January 1997, including the establishment a Pacific Resource Conservation Council by 1998, improved consultative mechanisms, and funding of $35.7 million. The funding is intended to support a variety of programs, including a habitat restoration and salmon enhancement pilot program, a gear payment program and an early retirement program.

Critics are skeptical, however, of the extent to which Ottawa will be willing to abandon the philosophy that led to the brink of disaster. Notwithstanding the announcement of the Pacific Salmon Revitalization Program, there have been further cuts of jobs and budgets to the tune of hundreds of millions of dollars in the federal Fisheries Department since 1994. Stop-gap bureaucratic measures, such as the tightening of enforcement and cuts to the commercial fleet, are seen by many as an inadequate response when the entire system is in a state of collapse.

Walters is particularly critical of the apparent unwillingness of bureaucrats to pursue new initiatives and opportunities suggested either by DFO staff or by outside interest groups. He cites several examples of proposals by fishermen to participate in programs for gathering better information for management and allocation. These initiatives described as "fundamental to successful fisheries management in the long-term," have received little or no support while enhancement and habitat restoration initiatives or examples of collaborative projects to grow fish in hatcheries are promoted. Walters calls for an "investment in information," not so much from the existing academic and government research establishment, but from fishermen and local communities because, he suggests while "science can tell us something about the limits of ecological performance..., only carefully acquired practical experience can tell us how to behave sustainably within those limits."[34]

B.C. fisheries have become highly capitalized and increasingly dependent on complex allocation arrangements during the current period of high marine productivity. Climate change experts suggest that it is unlikely that the very favorable environmental conditions responsible for this high productivity will last. As we enter a period of greater uncertainty the need for strongly adaptive management systems becomes more urgent. While the apparent recovery of sockeye stocks on the Fraser River in recent years

---

[32] To improve the financial performance of the fleet over time and moderate fishing pressure, a licence holder who wishes to fish in more than one area or with more than one type of gear must obtain an additional licence from another licence holder.

[33] The 1996 run of sockeye totalled 4.3 million fish, exceeding the precautionary pre-season forecast of 1.6 million, resulting in an escapement of approximately 2.2 million sockeye.

[34] C. Walters, *op. cit.*, 48.

suggests that the events of 1994 were less a crisis in fish stocks than a crisis in policy,[35] these crises cannot be repeated over and over without seriously threatening salmon resources. Examples such as the one outlined here call on us all to rethink the fundamental basis of how we manage our resources; that means seeing the cost of managing, the cost of protecting, the cost of enhancing the resource as an investment and not a cost.

Questions

1. With respect to the long-term institutional future of Canada's fisheries, would you advocate the maintenance of the current system of centralized authority and public responsibility for management or support a radical shift toward community-based management?

2. What factors encourage public agencies, such as the Department of Fisheries and Oceans, to engage more often in crisis management than in long term planning and management?

3. In what respect is the management of a migratory species, such as the sockeye salmon, more problematic than the management of a more sedentary population?

4. Artificial culture systems (such as hatcheries) have more modest habitat requirements and produce populations that can withstand much higher fishing pressure than wild populations. Discuss the merits and limitations of such approaches.

Further Reading

Marchak, P., Guppy, N. and J. McMullan 1987. *Uncommon Property: The Fishing and Fish-processing Industries in British Columbia.* Toronto: Methuen.

McGoodwin, J.R. 1990. *Crisis in the World's Fisheries: People, Problems, and Policies.* Stanford: Stanford University Press.

Meggs, G. 1991. *Salmon: The Decline of the British Columbia Fishery.* Vancouver: Douglas and McIntyre.

Newell, D. 1993. *Tangled Webs of History: Indians and the Law in Canada's Pacific Coast Fisheries.* Toronto: University of Toronto Press.

Pearse, P. H. 1992. *Managing Salmon in the Fraser: Report to the Minister of Fisheries and Oceans on the Fraser River Salmon Investigation.* Ottawa: DFO.

Pinkerton, E. 1989. *Cooperative Management of Local Fisheries: New Directions for Improved Management and Community Development.* Vancouver: University of British Columbia

Web Sites

- B.C. Ministry of Agriculture, Fisheries and Food: http://www.agf.gov.bc.ca/educate/tech.htm
- B.C. Ministry of Agriculture, Fisheries and Food: http://www.agf.gov.bc.ca/fish/aborigin/aaupdate.htm
- Canadian Stock Assessment Secretariat: http://csas.meds.dfo.ca/csas/index_e.htm
- Department of Fisheries and Oceans: http://www/ncr.dfo.ca/home_e.htm
- Pacific Salmon Alliance: http://diane.island.net/~psa/

Audio-Visual Material

- *British Columbia Salmon*, National Film Board, 1977 (dir.: Floyd Elliot).
- *For the Sake of Salmon*, Canadian Broadcasting Corporation, 1990 (30min.).

---

[35] P.H. Pearse, *Managing Salmon in the Fraser: Report to the Minister of Fisheries and Oceans on the Fraser River Salmon Investigation* (Ottawa: DFO, 1992), 29.

Case One

- *Life of the Sockeye Salmon*, Wilf Gray Productions, 1975 (24 min.).
- *Salmon People*, National Film Board, 1977 (dir.: Shelah Reljic and Peter Jones, 25 min.).
- *Salt Water People*, National Film Board, 1992 (dir.: Maurice Bulbulian, 119 min.).

Case Two
Great Whale: Lessons from a Power Struggle

Focus Concept

The interplay of knowledge and political power in public controversy over a proposed hydro-electric "megaproject," and the assessment of its environmental impacts.[1]

Introduction

The Great Whale Complex, also known as James Bay Phase II, called for the construction of a hydro-electric project with a capacity of 3,212 megawatts near Hudson Bay, 1150 km north of Montréal. The complex was to be located close to the adjacent villages of Kuujjuarapik, an Inuit community with a population of 460 and Whapmagoostui, a Cree village with a population of 500. The complex would require the construction of three hydro-electric generating stations, each with its own reservoir on the Great Whale River; the creation of a regulating reservoir at Lac Bienville; and the diversion of the waters of the Little Whale River. Scheduled for completion in 2010, these developments would involve the construction of eight dams and 128 dikes, the flooding of 3391 km^2, and reductions in the average annual flows of the Great Whale and Little Whale Rivers by 83% and 94%, respectively. The $12.6 billion project was shelved indefinitely by the Québec Premier, Jacques Parizeau, in November 1994. The chronology of events leading up to the cancellation of James Bay Phase II: Great Whale Complex is summarized in Table 2.1.

Power Struggles

Since the announcement in 1971 of the La Grande Complex (Phase I), the James Bay hydro-electric projects have been fought from every possible angle: in the courts, in the media and by diverse lobbies in national and international arenas. The controversy over the Great Whale Complex (Phase II)[2] provides a stark illustration of the fact that project evaluation is not reducible to the "neutral" and "objective" weighing up of scientific information as the privileged basis for determining public policy. The institutional process of environmental impact assessment, its goals, and the knowledge brought into play to fulfil them, are all shaped by power struggles between groups with differing, sometimes conflicting, values, ideologies and interests[3].

---

[1] For this case, I have borrowed extensively from lengthy discussions with Colin Scott, Dept. of Anthropology, McGill University, who also made available his research notes on the topic.

[2] Some observers refer instead to a second phase of hydro development of the *La Grande* watershed as 'Phase II.' This further development in the 'Phase I' watershed was larger in terms of generating capacity, and flooded nearly as much land, as the Great Whale Complex would have done. It occurred over the same period that the Québec Cree were opposing the Great Whale proposal, but 'La Grande Phase II' occasioned neither protracted resistance on the part of the Crees, nor meaningful environmental review (Alan Penn, personal communication). There are several possible reasons: 1. 'La Grande Phase II' did not inspire the support of the Crees' environmentalist allies; 2. It impacted watercourses already heavily damaged by 'La Grande Phase I;' and 3. Under the circumstances, by reaching a deal with Hydro-Québec not to oppose 'La Grande Phase II,' the Cree were making the best of a relatively weak position. The deal enabled regional leadership to make a significant concession to the pro-development minority within their own constituency, and solidify regional opposition to the Great Whale Complex.

[3] S. McCutcheon, *Electric Rivers: The Story of the James Bay Project* (Montreal/New York: Black Rose Books, 1991).

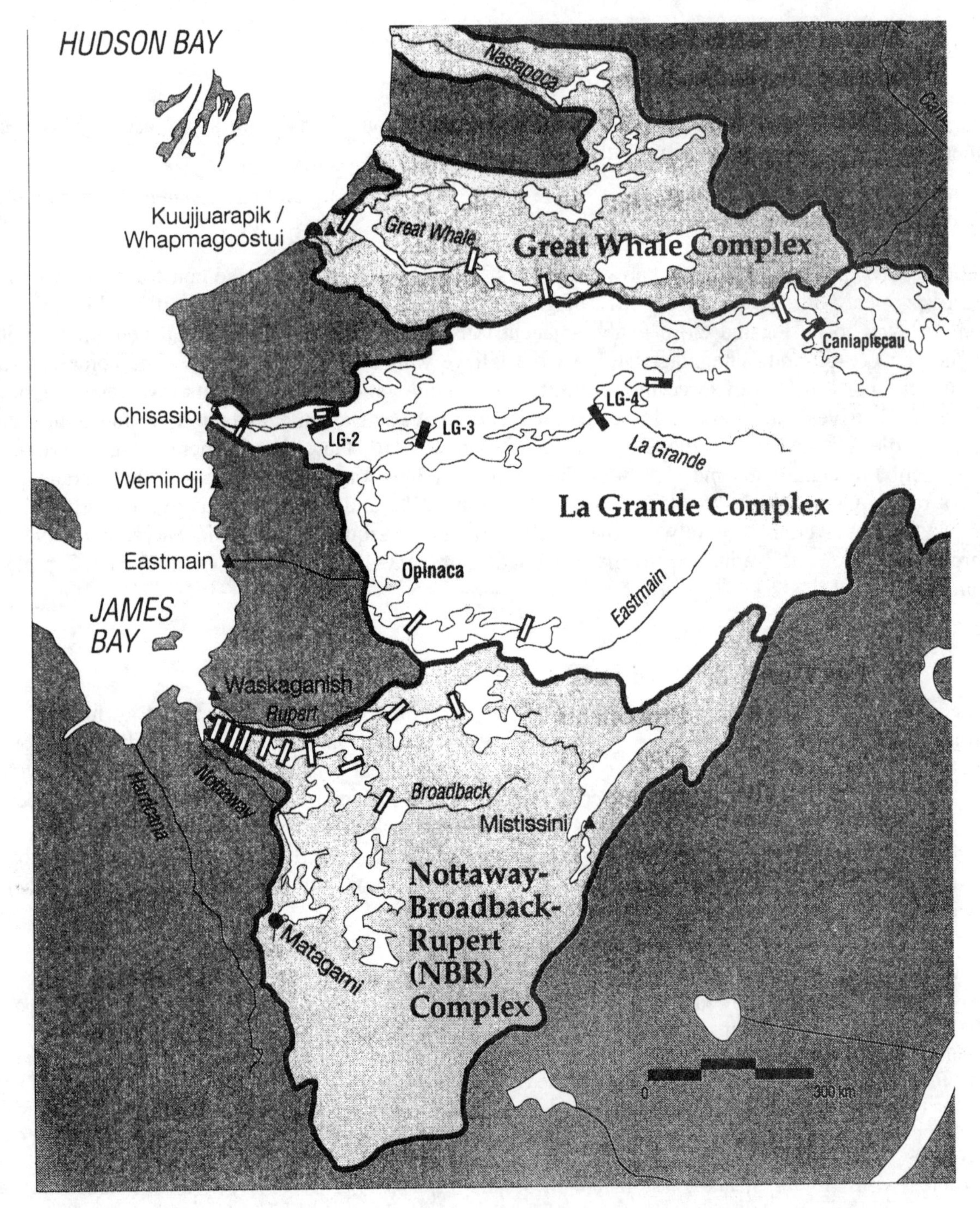

Source: S. McCutcheon, Electric Rivers (Montréal: Black Rose, 1991), ix; used with permission.

Figure 2.1. The James Bay Project

Hydro-Québec and the provincial government have always insisted that hydro-electric development is essential to Québec's economic and political future. According to their projections, the power generated by the Great Whale Complex would not only supply Québec's own needs but provide a significant export commodity when sold to the north-eastern states of the United States. They also suggest that hydro-electric power is environmentally preferable to alternative sources of electricity, such as fossil fuels or nuclear energy.

Environmental analysts and activists disagree. Their opposition to the Phase II is centered on three arguments: that the Great Whale Project would do further damage to the environment and hence to the rights and interests of aboriginal Cree and Inuit; that it makes uncertain economic sense; and that managing the demand for power (energy efficiency) is economically and environmentally preferable to building mega-projects.

Table 2.1. Chronology of Events – Great Whale Project

Year	Events
1971	• Québec Premier, Robert Bourassa announces decision to develop James Bay Phase I
1973	• Malouf decision imposes a temporary injunction to halt construction; Québec Court of Appeals overturns injunction one week later in favor of "balance of convenience"
1975	• James Bay and Northern Québec Agreement (JBNQA) signed by Inuit of Northern Québec, Cree of James Bay, Canadian government, Québec government, Hydro-Québec, the James Bay Energy Corporation and the James Bay Development Corporation
1985	• Construction of La Grande completed
1989	• Québec Premier, Robert Bourassa announces James Bay Phase II, Great Whale • Cree launch first of many law suits to block Phase II • Potential Hydro-Québec export contracts negotiated with north-eastern states • Ottawa announces that the project must meet federal environmental standards
1990	• Odeyak paddles into New York City Harbour, Earth Day • Cree file for a permanent injunction to block Great Whale Project • Conflict over splitting EA of access infrastructure from project • Hydro-Québec tables EIS of access infrastructure
1991	• New York City postpones final decision on 1000 megawatt contract • Federal and provincial governments agree to a joint environmental study of project • Rouleau decision forces signatories of JBNQA to abide by the Agreement and conduct • environmental impact hearings • Québec agrees to undertake a joint provincial-federal environmental study of project, including access infrastructure
1992 (Jan)	• Memorandum of Understanding between the federal and provincial governments and Cree and Inuit representatives to conduct a full environmental assessment subject to Sections 22 and 23 of the JBNQA
1992 (Mar)	• New York cancels 1000 megawatt contract with Hydro-Québec
1992 (Sep)	• Guidelines for EIS issued
1993 (Aug)	• Hydro-Québec releases its 30 volume ($400 million) feasibility study for the Great Whale project
1994 (Feb)	• Supreme Court of Canada rules that any power projects built to fulfil Hydro export contracts must undergo provincial and federal environmental reviews
1994 (Mar)	• New York Power Authority cancels 800 megawatt purchase agreement with Hydro-Québec
1994 (Nov)	• Hydro-Québec's 11-year, $256 million ($400 million after interest charges) impact study deficient in seven major areas and over 300 specific areas. • $13 billion, 3,212 megawatt Great Whale project shelved indefinitely by Québec Premier Jacques Parizeau

Native people living in the James Bay region are also opposed to the project. Despite their attempts to use the courts to stop construction of the La Grande Complex, Crees and Inuit eventually signed the James Bay and Northern Québec Agreement (JBNQA) in 1975, believing further opposition to be futile. In exchange for certain rights and benefits, including exclusive hunting, fishing and trapping rights to certain areas, a measure of self-government, and cash compensation, their aboriginal title to most of the land was extinguished. While Native communities and representatives acknowledge some benefits from the Agreement, they have experienced escalating social problems and economic challenges over the past two decades. These are attributed, in part, to the impacts of environmental destruction on their subsistence activities, and to procedures of metropolitan-based development and central government administration that conflict with local values and ways of life. These problems strengthened the resolve of Cree people throughout the region, and of local Inuit most directly affected by Phase II, to oppose it.

Today, the Cree and Inuit of Northern Québec are among the politically best organized native peoples in Canada. They formed alliances with environmentalists, economists, lawyers, academics and public relations experts in Canada and internationally to force the federal government to conduct one of the most ambitious environmental impact studies ever conducted for a major development project in Canada. The extent to which this review process addressed the range of perspectives and concerns of the various interest groups is examined below.

Political Analysis

It is helpful, in analysing political controversies, to ask ourselves the following questions: Who are the main interest groups involved? What are their goals and objectives? Their ideologies? What political resources do they command? And what are their strategies and tactics? Some background for answering these questions has been provided above. Let us briefly consider each in more detail.

Interest Groups and their Objectives

The interest groups may be divided into proponents and opponents of the project (see Table 2.2), with alliances forming between groups on either side.

The corporate goals of growth and profit made Hydro-Québec and its primary contractor, SNC-Lavalin, front-line proponents. They were supported by the Liberal cabinet of the Government of Québec, for whom the immediate burst of employment and business opportunities associated with the construction phase of such a large project could spell success at the polls in the next election. Investors sought profits, and unions sought jobs. Makivik Corporation, the regional political representative of the northern Québec Inuit, exploited the Québec Government's wish to divide the aboriginal opposition, and its public relations need to have an aboriginal group "on-side." Makivik negotiated a proposed half-billion dollar settlement, which they would have received in exchange for allowing the project to go ahead without resistance.[4]

---

[4] While local and regional self-governing bodies of the Cree united to oppose the project, local Inuit at Kuujjuarapik, who also opposed it, were not supported by their regional representative body, Makivik Corporation. The bulk of negative impacts of existing and proposed hydro development in northern Québec fall within the home lands and waters of the Cree communities. Only one or two Québec Inuit communities would share directly in the negative environmental impacts of Phase II, while the existing Phase I and the proposed Phase III fall entirely within Cree territory. Makivik Corporation, therefore, sought a deal with Hydro-Québec, deeming the potential benefits to be of sufficient overall value to its wider regional constituency to offset the hardship that would be experienced by Inuit locally.

The local Cree of Whapmagoostui and their regional government, the Grand Council of the Crees of Québec, were the lead opponents of Phase II, in alliance with a variety of environmental, human rights, industrial, and economic nationalist interests. Cree leadership sought to respond to broad-based anti-hydro sentiment among its people; to defend against the erosion of traditional resources and lifeways; and to demonstrate to the governments of Québec and Canada that development on traditional lands without Cree consent would be defeated. Their primary objective could be interpreted as Cree control of Cree lands and waters – not an anti-development position, but one that insists on forms of development that respect Cree autonomy, protect the environmental conditions of cultural continuity, and put local people first.[5]

Environmentalists, for their part, were motivated primarily by the goals of preserving wilderness and promoting environmental sustainability in energy policy; objectives that, in practical terms, could best be advanced by joining forces with the Cree.

Table 2.2. Interest Groups

Proponents:	**Opponents:**
Hydro-Québec	Cree Entities: - Grand Council of the Crees of Québec - Whapmagoostui Council
Major contracting and consulting companies - SNC-Lavalin	Inuit of Kuujjuarapik
Québec Government Cabinet	Indigenous rights NGOs - Cultural Survival (Canada)
Investment bankers	Environmental NGOs - Greenpeace - Friends of the Earth - Solidarity Foundation - Audobon Society - Sierra Club
Construction and service contractors	Public Policy NGOs - Coalition for a Public Debate on Energy - Canadian Arctic Resources Committee
Québec Federation of Labour	Energy Industry Competitors - coal mining - small hydroelectric plants - wood-burning, trash-burning co-generation companies
Inuit regional government (Makivik Corporation)	Miscellaneous - economic nationalists - anti-nuclear activists - bio-regionalists - eco-feminists - human rights activists - peaceniks

[5] It must be borne in mind that some Crees, even though in a minority, were attracted by the option of negotiating an agreement with Hydro-Québec on Great Whale. They anticipated cash economic benefits. On the other hand, many saw the point of demonstrating politically that development could not occur without Cree consent, and supported the opposition to Great Whale for this reason, even if not on environmental grounds.

Ideologies

The proponents share a similar ideology of progress through economic growth, achieved through harnessing the untapped potential of natural resources. In the words of the project's greatest advocate, former Québec Premier, Robert Bourassa:

> Québec is a vast hydro-electric plant in the bud, and every day millions of potential kilowatt-hours flow downhill and out to sea. What a waste! [6]

This ideology combines a "technological optimism with a utilitarian view of nature."[7] The economic benefits of hydro development are seen to outweigh any minor, local environmental damage, while the impact on Crees and Inuit is viewed as part of the inevitable modernization of natives.[8] Subsistence lifeways are thought to be withering away with the passing of elder generations. In any case, the (Québec or Canadian) national good is seen to transcend the rights and interests of indigenous "minorities".

Opponents of the project, of course, reflected a radically different blend of ideologies. Cree and Inuit emphasize their rights of self-determination as aboriginal peoples on traditional territories; the survival over millennia of cultures anchored in sustainable harvests from the land; their spiritual rapport with their traditional environments; and the superiority of their own knowledge of local and regional ecosystems.

Environmental activists argue that limits to capitalist growth must be found if we are to prevent the ultimate collapse of human ecosystems, and that mega-projects are models of how *not* to develop. Many reject as anthropocentric the premise that the last expanses of pristine wilderness should be harnessed to the needs of industrial society. Non-material values, including such "intangibles" as environmental aesthetics, are held to be superior to the crude material benefits of industrial growth. Many environmentalists have a romantic view of nature, and of indigenous peoples who are believed to co-exist harmoniously with nature.

Ideologies about the proper organization of the state and economic policy are also at stake. Decentralized decision-making, sensitive to the development needs of local and regional economies, and resulting in greater economic stability through diversity, is advocated over large-scale, centralized, hierarchical planning and decision-making.

Political Resources, Strategies and Tactics

The proponents had major financial and political resources, including official government backing, at their disposal. Provincial crown corporation status provides Hydro-Québec with many special advantages. The utility does not pay federal tax, and any debt incurred is unconditionally guaranteed by the provincial government – i.e., by the taxpaying public. Thus, while being the third largest company in Canada with assets of $36 billion, Hydro-Québec has a long-term debt of $26 billion, owed mostly to foreign banks. This translates as approximately $8,500 per Québecer, and means that half the corporation's gross revenues are spent servicing the debt. Yet, because of its political importance, most of its activities have not been subject to public review:

> Hydro-Québec forges ahead in the absence of any real accountability for its corporate behaviour. As a virtually unregulated monopoly, it successfully evades informed public

---

[6] R. Bourassa, *Power from the North* (Scarborough: Prentice-Hall, 1985), 4.
[7] S. McCutcheon, *op. cit.,* 148.
[8] *Op. cit.*, 147.

> scrutiny ... and is also immune to the discipline of the market, relying on government, guaranteed loans to finance its assault on the wilderness.[9]

The strategy for Phase II of Hydro-Québec, actively supported by the provincial government and with the acquiescence of the federal government, was to present the public with a *fait accompli*. This would be achieved, in part, by limiting and controlling the environmental review process. Jurisdictional stand-offs between the federal government in Ottawa and the provincial government in Québec gave the latter more scope than it might otherwise have enjoyed. Ottawa has been lax in its application of the federal environmental review process in recent years, not wishing to alienate provinces while it faces constitutional difficulties vis-à-vis Québec nationalism. The Cree had to resort to court action to force compliance of Québec and Canada with environmental law.

Hydro-Québec was able to deploy squadrons of lawyers, public relations experts, and other consultants on a scale unattainable by their aboriginal or environmentalist adversaries. These resources present huge opportunities for controlling scientific and public information. Financial means and governmental support, however, do not guarantee victory in battles for public opinion.

Opponents of the project, by contrast, included small but dedicated teams of local and regional indigenous leaders and their consultants, together with allies drawn chiefly from non-governmental agencies (NGOs) and universities. Proponents were usually young, highly motivated, well educated individuals with good access to networks of communication and support, including established international contacts. Notwithstanding inferior financial means, the enthusiasm and determination of the opponents, and their willingness to undertake punishing schedules and put in extremely long hours, were invaluable political resources.

Native parties were able to arrange comparatively modest but crucial financial support for researching and communicating their positions, thanks to public expenditures on the environmental assessment process, and by drawing on income from JBNQA compensation funds as well as personnel resources within their self-governing institutions. Certain NGOs received limited public monies as intervenors in the environmental assessment process, but depended primarily on volunteer inputs and private donors.

The legal rights of Cree and Inuit to substantial roles in environmental assessment, as defined by the JBNQA, and their organizational ability to press legal actions against reluctant central governments in defence of these rights, were key political resources. The Cree added years to the review of Phase II by implementing these rights. This success bought them time to mobilize support against Phase II in the northeastern United States, by simultaneously focusing attention on the project's environmental, economic, and human rights implications. The moral position of proponents, with its combined appeal to environmental integrity and human rights, was indispensable in the campaign for public sympathy.

Opponents generally went beyond the anti-colonist and anti-development rhetoric used by Natives and environmentalists in the past, to question the fundamentals of economic growth and to challenge current energy and political policies. Their approach was to present verbal arguments in a variety of fora, including energy regulatory bodies, environmental review panels, courts, legislative bodies, voters' assemblies, and media venues. Native activists launched a highly successful public-relations campaign, which included the voyage of the Odeyak, half canoe and half kayak, to represent the solidarity of local Crees and Inuit, respectively. Participants travelled from the villages of Whapmagoostui/Kuujjuarapik by dogsled, truck and water to arrive to a massive public rally in New York's Central Park on Earth Day, April 22, 1990.

---

[9] S. Hazell, Battling Hydro: Taming Québec's runaway corporate beast is a Herculean task. *Arctic Circle*, November/December, 1990, 41.

Although the Crees also threatened direct action, such as the blockade of roads and occupation of substations, their stated strategic preference was to avoid physical confrontation and to pursue legal action and public debate. In the end, the actions and arguments of the opponents took a heavy toll on financial and political support for the Great Whale project. The activism of the Cree and their environmentalist allies met with some success in encouraging public entities in the northeastern United States to divest themselves of Hydro-Québec bonds. More decisively, it contributed to the cancellation of purchase contracts for several billion dollars worth of electricity from Hydro-Québec. This fatally undermined the utility's ability to finance Phase II at acceptable levels of indebtedness.

Environmental Arguments

Proponents of the project claim that hydro-electric power is clean, unlike coal-fired alternatives; renewable, unlike fossil fuels; and safe, unlike nuclear fuels. In the words of Hydro-Québec's Chairman of the Board, Richard Drouin:

> The majority of the people of Québec believe in hydropower because it is a clean, safe, renewable and inexpensive source of energy. From our perspective it is clearly superior to the alternatives. That is why 95 percent of our electricity needs now come from hydropower. As a result, our greenhouse gas and acid rain emissions are relatively low. Indeed per capita emissions from energy sources are less than half as high in Québec as in the United States. Moreover, hydropower is more efficient than fossil-fuel energy.[10]

Environmentalists dispute the claim that hydro is one of the cleanest sources of electricity. While hydro-electricity produces energy without burning fossil fuels and is clean to the consumer at the end of the power line, there is much environmental damage at source: the loss of aquatic and terrestrial habitats, especially productive lowlands and waters, and the impact of these habitat alterations on migratory birds, caribou, beluga whales, rare freshwater seals and other wildlife in the James Bay territory. Phase II opponents cited the case of the 10,000 caribou drowned in 1984 while fording the Caniapiscau River, swollen by heavy precipitation and the release of water from an upstream reservoir. A report by the Government Secretariat for Indian Affairs concluded, "the management of Caniapiscau Reservoir was the principal cause of the drownings."[11] A Hydro-Québec report claims, however, that the drowning was "due to natural causes."[12]

Other problems relate to changes in water flow and sediment load. The average water level in Great Whale River would drop by more than 2 m, exposing shoals and the outer edges of the current river-bed to erosion.[13] The reduced flow would also result in changed ice conditions. Such modifications to both the flow and sedimentary regimes of the river would have inevitable downstream impacts on the marine environment, resulting in the disruption of currents, changes in water composition, and a different regime for the formation, movement and break-up of sea-ice. Such changes are likely to impose impacts on the biological productivity of estuaries and offshore eelgrass beds, which provide a vital habitat for the Canada goose and snow goose during their brief stopover each spring and fall in the shallow waters of James Bay.

Mercury contamination is perhaps the most contentious of the environmental issues. Reservoir flooding stimulates the production of poisonous methylmercury through a biochemical reaction that occurs when

---

[10] R. Drouin, Letter from Hydro-Québec. *The Amicus Journal*, Winter, 1993, 49.
[11] Cited in A. Gupta, R.F. Kennedy, and A.J. Scherr, Letter from NRDC. *The Amicus Journal*. Winter, 1993, 50.
[12] *Op. cit.*
[13] Hydro-Québec, Grande-Baleine Complex: Feasibility Study. Summary, August, 1993, 76.

plants decompose under water. The process of bioaccumulation concentrates mercury up the food chain, to be ingested by people who eat fish from contaminated waters. A 1984 survey of Cree living in the village of Chisasibi, an area affected by Phase I, found that 64% of the community had unsafe levels of mercury in their bodies. Hydro-Québec claims that opponents of the project overstate the risk of mercury and understate Hydro-Québec's efforts to mitigate the effects of mercury:

> Mercury levels in fish in northern Canada are *naturally* high, due in part to industrial and power plant pollution from the United States and Canada (my emphasis).[14]

Hydro-Québec claims that monitoring of fish and of people has brought the mercury problem under control in the short-term, and that mercury levels diminish over time and return to their original levels after 20 to 30 years. The National Resources Defense Council (NRDC) counters this claim, believing the exposure of people to mercury to be dangerously high and unlikely to abate for 80 to 90 years.[15] Mercury levels have dropped in recent years but at a price; Crees are warned not to eat fish from parts of the La Grande River and women of childbearing age are discouraged from eating fish from the area. Dietary losses of high quality protein, and other nutritional and health advantages of eating fish, are health impacts potentially as serious as the risk of contamination – not to mention the cultural impacts of disrupted subsistence practices.

Hydro-Québec's position on the global environmental impacts of the project is also at odds with the views of the environmentalists. While Hydro-Québec acknowledges that large bodies of water "produce microclimatic conditions within their immediate area of influence,"[16] they claim that the flooding of terrestrial habitats does not have continental or global scale environmental impacts; modifications to reservoirs "do not extend beyond the physical limits of the reservoir."[17] Environmentalists believe that not only does the cutting of trees in areas to be flooded reduce the amount of carbon dioxide extracted from the atmosphere, but the drowning of trees results in the release of carbon dioxide and methane, the most potent of greenhouse gases. Furthermore, the substantial surface areas of reservoirs increase the amount of water vapour released into the atmosphere.

Economic Arguments

Proponents of the project promote hydro-electricity as a cheap form of energy and a reliable investment, which is only slightly affected by inflation and avoids the uncertainties of fuel costs. Hydro-Québec also emphasizes the benefits to the wider economy: during phases of active construction, the utility pumps up to $2 billion per year into the provincial economy, and up to five dollars of every twenty dollars spent in Québec stem from Hydro-Québec spending. Hydro-Québec has 23,000 employees. An estimated 22,715 person-years would have been required for the construction of the Great Whale project,[18] touted as a superior means of job creation:

> Hydro-electric facilities generate the most jobs per million dollars of expenditure because they require major investments in civil engineering and construction, which are very labour-intensive.[19]

[14] R. Drouin, Letter from Hydro-Québec. *The Amicus Journal*, Winter, 1993, 49.
[15] A. Gupta *et al*, *op. cit.*
[16] Hydro-Québec, *op. cit.*, 74.
[17] *Op. cit.*, 250.
[18] *Op. cit.*, 48.
[19] *Op. cit.*, 10.

In addition to providing a boost for these "upstream" industries, Hydro-Québec also claims the project would provide significant spin-offs for "downstream" industries through lower rates charged to electricity intensive industries (e.g. aluminium smelters, magnesium refining). Furthermore, they propose "significant economic benefits for villages in the study area during construction of the project,"[20] including the claim that no other economic development project could justify the cost of providing the region with road access and other infra-structural improvements.

Opponents counter these arguments with details of the cost savings of various energy alternatives. Energy-saving programs and technologies are argued to be more efficient sources of permanent employment than short-term construction activity. Opponents are particularly critical of Hydro-Québec's level of public debt and suggest that this debt load will drive electricity rates higher. Concerns are also expressed about the creation of economic dependence. Over-investment in hydro-electricity is seen to have starved the provincial manufacturing sector of capital, leaving it weak and non-competitive. As a result, it is claimed that upstream benefits are limited, while the dependence of downstream industries on lower electricity rates requires that the power be supplied at a loss; secret risk-sharing deals with aluminium and magnesium producers are estimated to have already cost the province $2.9 billion over life of the contracts:[21] "Power projects and aluminium smelters are expensive and environmentally damaging ways to create relatively few jobs."[22]

With respect to the supposed benefits to Native communities, while substantial improvements in housing, education, health services, etc. have occurred over the past two decades, the rapid centralization of Crees and Inuit into structured communities that accompanied large-scale development decreased the proportion of people in the traditional economy. This situation has led to social instability in the villages that is reflected in high frequencies of such problems as unemployment, suicide, neglect of children, vandalism, and drug and alcohol abuse.[23] Meanwhile, hydro-electric operations offer few entrepreneurial or permanent employment opportunities to native residents of the region.

Supply-side versus Demand-side Arguments

Closely related to the economic arguments outlined above is the concept of demand management in energy conservation. McCutcheon refers to supply-side versus demand-side arguments as the "bulldozer coalition" versus " efficiency coalition" and sees the position of these two groups as critical to understanding the "pros" and "cons" of the project.[24]

Proponents of the project, in particular Hydro-Québec, emphasize energy supply. They claim that hydro-electric development is needed so that increased supply can keep pace with growth in demand. They justify their emphasis on supply-side management by claiming the utility would have to increase rates to consumers to pay for energy efficient programs (i.e. demand-side management).

Opponents point to Hydro-Québec's tendency, historically, to over-estimate future demand; and argue that much of actual demand is based on waste that could be eliminated. They suggest that higher rates to consumers to pay for energy efficient programs would be more than offset by lower energy consumption or *negawatts*. The term "negawatts" was introduced by energy analyst Amory Lovins, and refers to units

---

[20] *Op. cit.*, 263.

[21] R. Séguin, Secret pacts cost billions, Hydro-Québec studies show, *Globe and Mail*, March 31, 1993.

[22] H. Lajambe, as quoted in P. Gorrie, The James Bay Power Project: The environmental cost of reshaping the geography of northern Québec. *Canadian Geographic*, 110(1), 1990.

[23] R. Niezen, Power and dignity: The social consequences of hydro-electric development for the James Bay Cree. *Canadian Review of Sociology and Anthropology*, 30(4), 1993, 510-529.

[24] S. McCutcheon, *op. cit.*, 178.

of electric power which, if saved, the utility does not have to generate. They are cheaper than megawatts and create up to four times more jobs than hydro mega-projects for the same investment. Negawatt advocates also claim that this approach has a better multiplier effect for local economies because energy saving dollars recirculate more times than dollars spent on dams. "A good program of energy conservation would give us in megawatts approximately the equivalent of two Great Whale projects at one-fifth to one-tenth the cost" says economist Hélène Connor-Lajambe, founder of the Centre for Energy Policy Analysis in Montreal .[25]

Management of demand can also help avoid the social, environmental and economic costs of building new projects. In addition, environmentalists suggest that this strategy would increase the utility's flexibility by buying time to develop more benign forms of energy – solar, wind, biomass, geothermal and hydrogen. In other words, the outcome would mean conservation in the short term and less harmful sources in the long term.

Notwithstanding Hydro-Québec's claims to the contrary, a number of public reviews have indicated its lack of commitment to energy conservation and demand-side management programs. NRDC's recent review showed that, when compared to successful utility programs in other parts of Canada[26] and the United States,[27] Hydro-Québec's programs are poorly designed and under-funded, capturing only a fraction of the energy savings that are possible. Economists and energy analysts have repeatedly stated that "soft path" economics (i.e. conservation) make more economic sense than investing in massive new energy development. American energy economists at the World Energy Conference in Montreal in 1990 calculated that the north-eastern states would actually save money if they rejected James Bay power, paid for conservation programs in Québec, and then bought up the resulting energy surpluses. A 1984 energy report of the Harvard Business School and Energy Probe of Canada also advocated conservation as the key to meeting future energy demands. The study, entitled "Soft Energy Futures for Canada," showed that with no additional investment in electricity, conservation measures at home could allow Québec to extensively export electricity and still meet its domestic energy needs. Despite these findings, Hydro-Québec maintains its commitment to supply-side management; a shift to demand-side management may be unthinkable because it would require the severing of substantial benefits to other politically influential proponents.

The Great Whale Environmental Public Review

The environmental review processes for the Great Whale project were framed by a Memorandum of Understanding (MOU), signed in January 1992 by the governments of Canada and Québec and representatives of the Cree and Inuit of Québec. The Memorandum provides for the co-ordination of the review procedures as laid out in the JBNQA, as well as the federal Environmental Assessment and Review Process (EARP).

The co-operation of the governments of Québec and Canada was gained only as the result of numerous lawsuits launched by the Cree. The provincial government had originally attempted to separate the assessment of required roads and other access infrastructure from the assessment of the hydro-electric installations. Québec argued that the access infrastructure belonged in a category of lesser impact than the

[25] H. Lajambe, *op. cit.*

[26] In B.C., the introduction of the Power Smart Program will by the year 2000 have saved B.C. hydro the amount of power previously consumed by 400,000 homes in the province.

[27] In the United States, the Bonneville Power Utility spent $900 million on conservation programs in the 1980s and postponed any new large-scale energy generation projects indefinitely. Consolidated Edison, a major utility that supplies New York city, will save over $1 billion annually over the next two decades by reducing its demand by 21.8%.

hydro installations *per se*, and should not involve federal jurisdiction. The provincial government hoped that it could grant early clearance for the access infrastructure, so that construction could proceed while the more complicated review of the dams and reservoirs was completed. Not to proceed with the hydro installations after massive public expenditure on transportation infrastructure would then have become politically unthinkable. Hydro-Québec and the Government of Québec could then have presented the project as a *fait accompli.*[28] The federal government, anxious not to anger Québec, did not object. The Crees argued that this strategy was unacceptable from both legal and ecological perspectives. They won a declaration from the Québec Superior Court in 1991 that the division of the mega-project into two components for the purposes of assessing environmental impacts was illegal.

Review Bodies and Process

Six different environmental assessment and review bodies were involved in the Great Whale Project. Five of these were established by the JBNQA, while the sixth was established by the Federal EARP:

- Kativik Environmental Quality Commission (KEQC), a provincial panel that examines environmental, social and economic impacts north of the 55th parallel;
- Federal Review Committee (COFEX-NORD), which examines the impact of projects north of the 55th parallel;
- Evaluating Committee (COMEV), a federal-provincial body that issues directives on impact studies for work south of the 55th parallel;
- Provincial Review Committee (COMEX) that examines impact studies for projects south of the 55th parallel;
- Federal Review Committee (COFEX-SUD), which examines the impact of projects south of the 55th parallel; and
- Federal Environmental Assessment Review Panel (FEARP), which examines environmental impacts in areas under federal jurisdiction, such as fisheries and marine mammals.

The review process involved the following steps:[29]

1. Proponent announces its intent to develop
2. Public "scoping" hearings provide input to study guidelines
3. Review bodies issue study guidelines
4. Proponent conducts study
5. Review bodies assess conformity of study to guidelines
6. Remedy of deficiencies in the study
7. Acceptance of revised study by review bodies
8. Public hearings
9. Review bodies deliberate and recommend

While the review bodies had different compositions, timetables and legal mandates, the MOU provided for their harmonization in the overall process.

Guidelines for the Review

The guidelines for the environmental impact statement (EIS),[30] issued in September 1992, required Hydro-Québec to demonstrate the need for new or additional electrical generating capacity and energy

[28] This stratagem caused dissension within the ranks of cabinet. The objections of the Minister of Environment, Pierre Paradis, however, were swept aside by the Minister of Energy, Lise Bacon, and Premier Robert Bourassa.
[29] In the event, the review reached only step 5 in the process before the Great Whale Project was called off.

resources, and to examine all possible alternatives for meeting the anticipated growth in demand, including energy efficiency and demand-side programs. The guidelines also specified that:

- evaluation of the proposed project be based upon the five fundamental issues that affect *ecosystem integrity*: 1) the health of all human and animal communities in the region, 2) access to the territory by all human and animal populations, 3) the availability of resources, which requires that they be sufficient in quantity and satisfactory in quality, 4) the maintenance of social cohesion at the local, regional and national levels, and 5) the respect for values;
- Hydro-Québec establish mechanisms to incorporate the knowledge and opinions of all communities that inhabit the region. Since points of view concerning the proposed project and its impacts are based on different values and knowledge, Hydro-Québec must take this diversity into account by defining the environment in accordance with state-of-the-art scientific methods, as well as in accordance with the acquired knowledge of the Crees and Inuit. The notion of *valued ecosystem components* will be applied in providing this multicultural definition of the environment;
- the EIS address the combined and *cumulative effects* of impact on entire sectors of the ecosystem, including the human societies in the proposed project area; and
- Hydro-Québec address potential *trans-boundary impacts*, e.g. on migratory birds, and on global climatic conditions through the production of greenhouse gases in areas that would be completely or partially submerged.

Theory versus Practice

The guidelines for the Environmental Impact Statement (EIS) are the most ambitious ever issued for a development project in Canada. They comprise 95 pages, 373 requirements, and embrace concepts of sustainable development, ecosystem integrity, cumulative impacts, and trans-boundary effects. They call for a reconceptualization of ecosystems to include human activity, and require that cultural relativism be employed in evaluating environmental impacts.[31] As a result, the proponent was required to meet the challenge of integrating mainstream western "science" and local knowledge of socio-environmental impacts within the framework of environmental assessment.

The degree to which this challenge went unfulfilled is reflected in the disparity of expenditure on the scientific research of the proponent versus documenting the knowledge of local Native people. A total of $450,000 and seven months in 1994 were devoted to the community consultation study, as compared with $256 million ($400 million after interest charges) that Hydro-Québec claims to have spent over eleven years of feasibility and assessment research.[32]

Furthermore, there was no serious agenda for pursuing scientific research into many of the impacts identified within Cree local knowledge. To illustrate the unreconciled divergence in understandings of environmental and social impacts, the conclusions put forward by Hydro-Québec in the Great Whale

---

[30] Guidelines: Environmental Impact Statement for the Proposed Great Whale River Hydro-electric Project, Evaluating Committee, Kativik Environmental Quality Commission, Federal Review Committee North of the 55th Parallel, Federal Environmental Assessment Review Panel, 1992.

[31] Notwithstanding a concession gained by Québec under the JBNQA, with regards to future hydroelectric development that "sociological impacts shall not be grounds for the Cree and/or Inuit to oppose or prevent said developments" (James Bay and Northern Québec Agreement, Éditeur Officiel du Québec, 1975: 111) the Guidelines recognize that because environmental impacts are culturally and socially defined, social and cultural factors could not be excluded from environmental assessment.

[32] This latter figure should be taken with a grain of salt; Hydro-Québec is accused by some informed observers of grossly exaggerating its expenditures on environmental research (Alan Penn, personal communication).

Impact Statement are summarized in Table 2.3, in juxtaposition to the observations and interpretations of local Cree provided during the community consultations.[33]

In November 1994, the Commissions and Committees charged with the review of the conformity and quality of the EIS found the study lacking in seven major areas and required that 300 specific deficiencies be addressed. The proponent never complied, because within days a newly elected Québec government announced that it was shelving the Great Whale Project indefinitely.

Table 2.3. Divergent Perspectives on Hydro-Electric Impacts

Hydro-Québec's view	Local Cree view
Environmental Impacts:	
Vastness and relative homogeneity of environment	Environment as a patchwork of diverse micro-habitats
Only 0.2% of Québec's territory would be modified	0.2% represents only areas actually flooded or otherwise physically transformed; excludes other modifications (e.g. altered salinity in estuaries, impact on regional wildlife populations)
New environment created would be more productive and more stable, particularly reservoir areas; modifications to lakes and water courses would only affect a negligible proportion of animal populations	New environment would be less predictable, more dangerous and largely contaminated; major long-term impacts on regional wildlife populations
Little change in water quality, which always remains suitable to primary production and wildlife in general	Changes in water quality associated with the reduction or disappearance of several valuable species (e.g. sturgeon); or their inedibility
Riparian vegetation (i.e. along river banks) re-establishes over 10 years	Productivity of diverted rivers permanently reduced; flow regimes along remaining water courses largely prohibit stable re-vegetation
Mercury levels would render some of the biomass (i.e. 100% of piscivorous (fish-eating) fish and 20% of non-piscivorous) unfit for regular consumption during 30 years, but available biomass would be greater than before the project	Fish biomass largely unharvestable or culturally unfit for consumption; reduction in migratory species habitat
Migratory waterfowl depend on coastal habitats and would not be affected by inland flooding	Waterfowl have changed their migration habits in response to presence of large inland reservoirs
Social Impacts:	
Integration into modern society has already caused significant social upheaval	More social upheaval is not the answer
Hunting-fishing-trapping no longer ensure survival, nor even livelihood	There are as many Cree hunters and as much Cree hunting effort today as there has ever been
Presence of new road system provides for more rapid and frequent access to the community's inland hunting territory and fosters a more even distribution of inland wildlife harvesting	Roads have increased number of sport hunters in the territory causing caribou to move beyond range of local Cree hunters; sport hunting along roads also conflicts with the indigenous tenure system and important cultural values

[33] Cree views are summarized from research conducted by C. Scott and K. Ettenger (personal communication) for the Great Whale environmental assessment community consultation, Grand Council of the Crees of Québec/ Cree Regional Authority, 1994.

Lessons Learned

Public decisions about large-scale development projects, the identification of environmental and social impacts, and the balancing of the pluses and minuses, take place in highly politicized settings. Social actors in these settings frequently possess divergent interests, goals and values, and unequal means for achieving desired outcomes. The disparities are perhaps especially marked when culturally distinctive groups are involved, but may be expected wherever people with strong ties to a local region confront the allied interests of external capital and central governments. For the latter, local environments may be viewed primarily as repositories of resources to be extracted, and the deck seems stacked against long-term residents who have a very different vision of what is needed to maintain or improve conditions in their communities and home areas.

Western science represents no monopoly on accurate information about existing and potential impacts of industrial development. All knowledge is generated in a particular social context, so that certain important questions are possible or even obvious, while other important questions may simply not be asked. This is not to deny that some knowledge may be more accurate or comprehensive than other knowledge, or that knowledge can be improved. But it *is* to argue that all knowledge is limited by cultural and historical parameters that focus it on particular social preoccupations. Cree and Inuit know certain things about the environmental impacts of hydro-electric development that escape science, because those impacts matter to them in ways that are normally far from the concerns of central government managers or professional scientists. Moreover, Cree and Inuit have opportunities for observing, and possess paradigms for interpreting environmental processes that differ quite radically from those of urban intellectuals.

Under such conditions, *which* knowledge becomes the basis for policy decisions, and *whose* social goals and values will be supported by that knowledge, will be heavily influenced by the relative power of competing groups. The degree to which local knowledge gets a serious hearing, and the degree to which local people can arrange scientific research to respond to their own priorities and agendas, reveals a good deal about the quality of our participatory democratic institutions. In view of the success of many indigenous societies in achieving long-term sustainable adaptations, the extent to which the wider society accommodates their knowledge may also reveal a good deal about our capacity for environmental health. Multiple sources of knowledge, moreover, provide for greater flexibility and reliability than fewer sources of knowledge. The guidelines of the Great Whale assessment (in concept if not in execution) suggest that that local knowledge, natural science and social science may be brought into more regular interaction in the appraisal of major development projects in Canada, and in the formation of environmental and social policy more generally. A substantial body of literature is emerging to assist in meeting this challenge.[33]

It is clear that these objectives will not be achieved without struggle. The Great Whale episode offers some hope that the structures of big government and the expansionary designs of metropolitan capital are not unassailable. Issues of social justice, environmental integrity, and human rights, in an age of transnational media awareness, can generate powerful movements that leave institutional landscapes transformed. Some observers argue, not without merit, that economic considerations were the decisive factors in the demise of the Great Whale project – that other ways of meeting energy needs in the northeastern United States became more competitive with the passage of years, and more attractive to investors. In the absence of the campaign led by the Cree, however, domestic and international scrutiny of

[33] M.M.R. Freeman and L.N. Carbyn, *Traditional Knowledge and Renewable Resource Management in Northern Regions* (Edmonton: University of Alberta, 1988); J.T. Inglis, *Traditional Ecological Knowledge, Concepts and Cases* (Ottawa: International Program on Traditional Ecological Knowledge and International Development Research Centre, 1993); and J. Mailhot, *Traditional Ecological Knowledge: The Diversity of Knowledge Systems and their Study*, Background Paper No. 4 (Montreal: Great Whale Environmental Assessment, Great Whale Public Review Support Office, 1993).

the project would have been more cursory, the alternatives would have been considered less seriously, and Phase II construction could well have gone beyond the point of no return by the time the Québec Liberals were voted out of power in 1994.[34]

Questions

1. To what extent is the interest of 6.5 million Québecers in hydro resources reconcileable with the interest of 10,000 Crees in their traditional homeland?

2. What is the potential, and what are the obstacles, for the complementary use of indigenous knowledge and scientific research in environmental impact assessment?

3. How stable is the alliance between environmentalists and native people? Where do the interests and ideologies of these two groups coincide, and where do they diverge?

Further Reading

Gedicks, A. 1994. *The New Resource Wars: Native and Environmental Struggles against Multinational Corporations*. Montreal: Black Rose Books.

McCutcheon, S. 1991. *Electric Rivers: The Story of the James Bay Project*. Montreal/New York: Black Rose Books.

Posluns, M. 1993. *Voices from the Odeyak*. Toronto: NC Press Limited.

Richardson, B. 1975/1991. *Strangers Devour the Land: The Cree Hunters of the James Bay Area vs. Premier Bourassa and the James Bay Development Corporation*. Toronto: MacMillan of Canada.

Salisbury, R. 1986. *A Homeland for the Cree: Regional Development in James Bay*, 1971-1981. Montreal: McGill-Queen's University Press.

Web Sites

- Grand Council of the Cree: JBNQA: http://www.gcc.ca/jbnqa/
- Grand Council of the Cree, Main menu: http://www.gcc.ca/mainmenu.htm
- Hydro-Québec: http://www.hydro.qc.ca
- James Bay Cree and Québec hydro Development: http://www.lib.uconn.edu/arcticcircle/culturalviability/cree/feit/readings.html
- James Bay Cree, sociocultural change: http://www.lib.uconn.edu/arcticcircle/historyculture/cree/creeexhibit.html
- Sierra Club: http://202.242.165.13/Env/Env.Links/ftr-canada.html

Audio-Visual Material

- *The Dammed,* Canadian Broadcasting Corporation, 1994 (92 min.).
- *Energy Québec*, Vidéo Québec – government of Québec, 1987 (13 min.).

[34] In the days before this text went to press, Hydro-Québec announced a proposal to divert upstream portions of the Great Whale and Nottaway-Broadback-Rupert (NBR) watersheds, via major canals, into the existing La Grande system of reservoirs and turbines. This would supplant Hydro-Québec's earlier long-term program to develop Great Whale (Phase II) and NBR (Phase III) independently. The new proposal follows on the heels of public undertakings by the Premier of Québec to achieve revenue-sharing agreements with aboriginal people on resources extracted from their traditional homelands. Sub-normal precipitation in recent years, resulting in the lowering of water levels in the utility's reservoirs to about one-fifth of capacity province-wide, has strained Hydro-Québec's ability to meet its domestic and export commitments. The Crees are revisiting, in this modified context, the choices between cash economic growth through capturing a share of the benefits of hydro development, and safeguarding other environmental and cultural priorities.

- *Flooding Job's Garden*, National Film Board of Canada, 1991 (dir.: Boyce Richardson, 57 min).
- *Hydro Québec-Our Natural Resources*, Hydro Québec, 1991 (16 min.).
- *James Bay: the Wind that Keeps on Blowing*, Canadian Broadcasting Corporation, 1990 (96 min.).
- *Power*, Cineflix, Productions Inc., 1996 (dir.: Magnus Isacsson, 77 min.).
- *Power of the North*, A Wild Heart Production, 1994 (dir.: Catherine Bainbridge and Anna Van der Wee).
- *Riding the Great Whale (Sur le dos de la Grande-Baleine)*, National Film Board of Canada, 1994 (dir.: D. Beaudry, 57 min.).

Case Three
Zebra Mussels: Environmental Threat or Aquatic Nuisance?

Focus Concept

The present and potential impacts of an exotic species on the Great Lakes ecosystem.

Introduction

"Exotic" species – organisms introduced into habitats to which they are not native – are world-wide agents of habitat alteration and destruction. They have become a major cause of biological diversity loss throughout the world and are often considered "biological pollutants".[1]

The introduction of a species either accidentally or intentionally, from one habitat to another, is risky business. Often, freed from the predators, parasites, pathogens, and competitors that have kept their numbers in check, species introduced into new habitats overrun their new home and crowd out native species. In the presence of enough food and a favorable environment, their numbers explode. Once established, exotics can rarely be eliminated. Some exotic introductions are ecologically harmless and some are even beneficial. But many exotic introductions are harmful to industry, recreation and ecosystems. They may cause the extinction of native species – especially those of confined habitats such as islands and aquatic ecosystems.

Most species introductions are the work of humans. Some introductions, such as carp (*Cyprinus carpio*) and purple loosestrife (*Lythrum salicaria*), are intentional and do unexpected damage. But many exotic introductions are accidental. These species are carried on animals, vehicles, ships, commercial goods, produce and clothing. The recent development of fast ocean freighters has greatly increased the risk of new exotics in the Great Lakes region. Ships take on ballast water in Europe for stability during the ocean crossing. This water is pumped out when the ships pick up their loads in Great Lakes ports. Because the ships make the crossing so much faster now, and harbors are often less polluted, more exotic species are likely to survive the journey and thrive in new waters.[2]

A variety of aquatic exotic plants and animals arrived to the Great Lakes in this way. They are now being spread throughout the continent's interior in and on boats and other recreational watercraft and equipment. Among the most problematic of these exotics are the ruffe (*Gymnocephalus cernuus*), common carp (*Cyprinus carpio*), zebra mussel (*Dreissena polymorpha*), spiny water flea (*Bythotrephes cederstroemi*), Eurasian watermilfoil (*Myriophyllum spicatum*), and purple loosestrife (*Lythrum salicaria*).

This case study focuses on the zebra mussel, specifically in terms of its biology, impacts and control. Less than a decade ago only a handful of people in North America had heard about zebra mussels. Today, zebra mussels are established in all the Great Lakes and several inland waters, and have become a household word in the Great Lakes region.

---

[1] Minnesota Department of Natural Resources, A field guide to aquatic exotic plants and animals, 1993.
[2] *Op. cit.*

Biological Attributes

The zebra mussel (*Dreissena polymorpha*) is a small freshwater bivalve mollusk. The shell, as the name *Dreissena polymorpha* implies, is highly variable in form and exhibits a dull, highly variable "zebra" pattern of irregular brown and cream concentric bands. Zebra mussels can live for five years and grow to about five centimeters in length, but most live for two to three years and are less than three centimeters long.

Zebra mussels have a planktonic larval stage, similar to most marine bivalves. However, this mode of reproduction is exceptional for freshwater bivalves. Females normally mature in their second year, but accelerated growth in the Great Lakes can produce mature females by the end of their first summer. Fully mature female mussels are reported to produce 30,000-40,000 eggs per season, although some scientists suggest that fecundity figures can be closer to 1 million eggs per season.[3]

Conditions needed for the successful development of the embryos include a temperature between 12° and 24°C (optimum at about 18°C), pH between 7.0 and 9.0 (optimum at about pH 8.5), and calcium levels above 20 milligrams per litre. Egg production starts when the water temperature warms to about 12°C, usually in early May in the Great Lakes. Eggs are fertilized outside the shell and within a few days hatch into microscopic free-swimming larvae, known as vilegers. Zebra mussels continue producing eggs until the water cools below 12°C, generally in October. The free-swimming veligers can remain suspended in water for about two to three weeks, during which time they are transported great distances by currents. Once their shells begin to form they become too heavy to float, and settle upon any hard surface. There they transform themselves into the typical, double-shelled mussel shape, which is clearly visible to the human eye. A tuft of small proteinaceous fibres, known as byssal threads, is later generated from a gland in their feet (known as the byssal gland). These threads attach to hard surfaces with an adhesive secretion that anchors the mussel in place.

Larvae are most vulnerable during the process of settlement if a suitable substrate cannot be found or if strong water movements prevent their settlement and impede their feeding due to a significant amount of particles in suspension. Larvae tend to settle at depths between 0.5 and 4.5 m, with a preference for greater depths later in the season. While mud and sand are unsuitable substrates, the substrate where larvae can settle is not very specific. The color and structure (rough or smooth) of the substrate seem to be unimportant, although larvae prefer to settle on the underside of objects in a location with a certain amount of water current. When necessary, young mussels can break away from their attachments and generate new, buoyant threads that allow them again to drift in the currents and find a new surface. Zebra mussels will attach themselves to any firm surface such as rock, gravel, metal, wood, vinyl, glass, rubber, crayfish, native mussels, and to each other. They will also attach to softer surfaces such as aquatic plants. Beds of mussels in some areas of the Great Lakes (parts of western Lake Erie) are approaching densities of $300{,}000/m^2$.

Although zebra mussels are freshwater species, there is some evidence that they can also survive in estuarine and coastal environments. Zebra mussels have been identified in the lower tidal reaches of the St. Lawrence River, but preliminary research findings indicate an upper limit of salinity to which the mussels can adapt. Out of water, zebra mussels are very hardy. With their shells closed, they can survive drying for several days. In moist environments, they can survive out of water even longer; adult zebra mussels are known to live from several days to two weeks in moist, shaded areas, such as bilges, live

---

[3] M. Sprung, The other life: An account of present knowledge of the larval phase of *Dreissena polymorpha*, in T.F. Nalepa and D.W. Schloesser, eds., *Zebra Mussels: Biology, Impacts, and Control* (Boca Raton: Lewis Publishers, 1993), 39-53.

wells, and the inside of trailer frames. This gives them the ability to invade new habitats that are not connected to infested waterbodies.

Introduction to North America

The zebra mussel formed as a species millions of years ago in the huge saline water basin which includes the present Caspian, Aral and Black Seas. It was at the mouth of the rivers that flowed into these seas that a stable relationship between the mussel and other organisms established.

About 200 years ago, evidently in connection with the construction of canals and increased euthrophication of the water basins, the zebra mussel began to increase its range into the North European rivers that emptied into the Baltic Sea. Zebra mussels were first found in the lower course of the Ural River in 1769. They became of great interest during the 1820s when they were found at the London docks and then in different places of western Europe. In Germany they acquired the name of the wandering mussel ("Wundermuschel") because of their ability to spread rapidly to different areas. Later in the 19th century the mussel began to block water supply pipes in Paris, Berlin and many other towns and cities throughout Europe. Suitable habitat increased considerably during the 20th century as a result of the conversion of several estuaries into freshwater lakes by damming them from the sea and the flooding of extensive regions as a result of the construction of reservoirs for hydroelectric stations.

Zebra mussels were first discovered in North America in late 1986.[4] It is generally believed that the mussel was introduced from Eastern Europe as "an unknown stowaway" in the form of larvae when one or more transoceanic ships discharged ballast water into Lake St Clair, near Detroit, Michigan. Another possibility suggested by some scientists, though not widely accepted, is that adult zebra mussels were able to attach themselves to the ship's anchor or coils of rope that remained moist enough during the voyage for the mussels to survive. There is consensus that the mussels most likely came from the warmest portion of the zebra mussel's European range, along the northern shore of the Black Sea.

Tolerant of a wide range of environmental conditions, within a few years of the arrival of the first individuals, zebra mussels had spread to all of the Great Lakes.[5] The entire Canadian shorelines of lakes St. Clair, Erie, and Ontario, the St. Lawrence River, most of Lake Huron, and portions of Georgian Bay and the North Channel are now colonized by zebra mussels (Figure 3.1). No well-established colonies have been found in Lake Superior as a likely result of low temperatures and reduced calcium levels. Veligers were detected in the Kawartha Lakes chain, Lake Simcoe, and the Muskoka Lakes in 1991, marking the invasion of zebra mussels into Ontario's inland waters. Zebra mussels have since colonized Rice Lake, and portions of the Trent Canal and Lake Simcoe; however, substantial colonies of mussels are not expected to develop in the Muskoka Lakes due to low calcium levels. In 1993, zebra mussel colonies were detected at several locations in the Rideau Canal system, which connects eastern Lake Ontario to the Ottawa River at Ottawa. Populations in this area continue to expand. In 1994-95, zebra mussels were discovered in six eastern Ontario lakes, which are not connected to infested waterbodies. In the United States, zebra mussels have colonized most of the Mississippi River, several of its tributaries, and many inland lakes.

---

[4] R.W. Griffiths, D.W. Schloesser, J.H. Leach, and W.P. Kovalak, Distribution and dispersal of the zebra mussel (*Dreissena polymorpha*) in the Great Lakes Region, *Can. J. Fish. Aquat. Sci.*, 48, 1991, 1381-1388.

[5] Details of the dispersal and current distribution of zebra mussels have been taken from Krishka *et al.* Impacts of Introductions and Removals on Ontario Percid Communities, Percid Community Synthesis, Introductions and Removals Working Group, Ont. Min. Nat. Resource, Peterborough, 1996, 41.

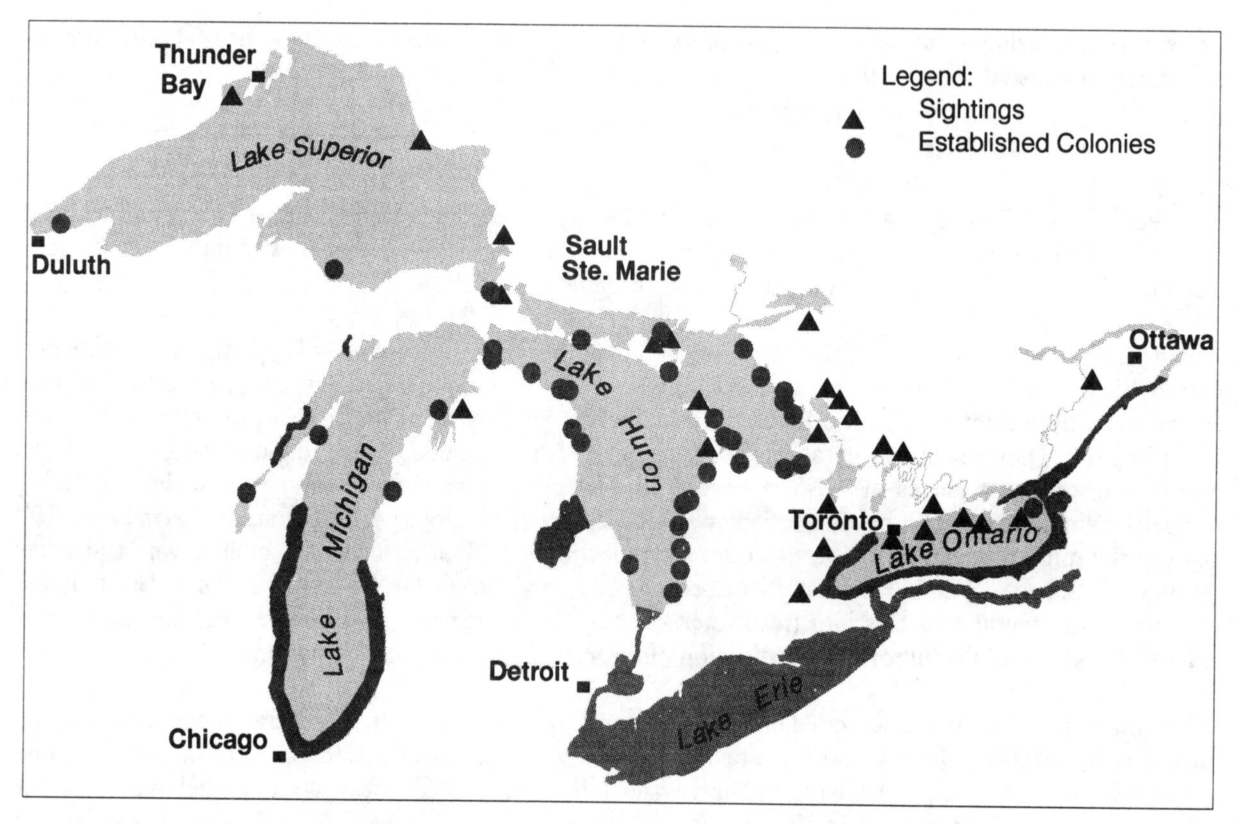

Source: Ministry of Natural Resources, Ontario

Figure 3.1. Zebra Mussel Distribution in 1996

The increase in the numbers of zebra mussel in the Great Lakes region in the first few years after its introduction far exceeds increases noted in Europe. This rapid advance has been facilitated by its high fecundity, its planktonic veliger larva, and a scarcity of natural predators. Despite their record high numbers, early indications suggest that mussel populations vary widely depending on the type of lake and habitat into which they spread. The main factors controlling the success of zebra mussel colonization in the North American waterbodies seem to be temperature, pH, and calcium levels. Waterbodies with a pH greater than 7.4, temperatures above 12°C, and calcium levels greater than 20 milligrams per litre are most likely to be colonized. The potential spread of zebra mussels in Ontario has been mapped using these factors as well as provincial transportation routes. Based on these maps, most waterbodies south of the Precambrian Shield are vulnerable to colonization, while shield lakes are unlikely to be colonized because of reduced pH and calcium levels.

Dispersal

The zebra mussel has a high reproductive rate, a microscopic free-floating larval stage, and the ability to attach itself firmly to any hard surface. These are essential to its rapid dispersal rate. The dispersal of the zebra mussel is mediated by three natural and twenty human-related mechanisms (Table 3.1).[6] Natural mechanisms include currents, birds, and other animals. Human mechanisms include those related to waterways, vessels, navigation, and fishery activities, and a wide variety of miscellaneous vectors (including intentional movements, aquarium releases and scientific research). All mechanisms transport

[6] J.T. Carlton, Dispersal mechanisms of the zebra mussel (*Dreissena polymorpha*), in T. F. Nalepa, and D.W. Schloesser, eds., *op. cit.,* 677-697.

juveniles and/or adults but fewer mechanisms transport eggs and larvae (e.g. currents, perhaps animals, canals, ballast water, other vessel water, fish stocking, bait bucket, firetruck water, aquarium releases, amphibious planes and scientific research).

Table 3.1. Potential Dispersal Mechanisms of Different Life Stages of the Zebra Mussel (*Dreissena polymorpha*)

		Life Stage		
		Planktonic	Sedentary	
Dispersal Mechanisms		**Eggs and Larvae**	**Juveniles**	**Adults**
Natural				
1	Currents[a]	X	X	X
2	Birds[b]	X?	X	X
3	Other aniimals	X?	X	X
Human-Mediated				
Waterways:				
4	Canals[a;] vessels, irrigation	X	X	X
Vessels/Navigation				
5	Ballast water[b]	X	X	
6	Vessels exteriors[a]		X	X
7	Vessels interiors	X	X	
8	Fishing vessel wells	X	X	
9	Navigation buoys[c]		X	X
10	Marine/boatyard equipment[c]		X	X
Fisheries:				
11	Fishing equipment		X	X
12	Fish cages[c]		X	X
13	Fish stocking water	X	X	
14	Bait and bait-bucket water	X	X	X
Other industry:				
15	Commercial products[a]		X	X
16	Marker buoys and floats[c]		X	X
17	Firetruck water	X	X	X
Other:				
18	Intentional movements[a]		X	X
19	Aquarium releases[a]	X	X	X
20	Amphibious planes	X?	X	X
21	Recreational equipment		X	X
22	Litter (garbage)[c]		X	X
23	Scientific research	X	X	X

Note: ? = Uncertain whether this life stage would be associated with the indicated mechanism.

[a] Dispersal by this mechanism has been documented.
[b] Dispersal by this mechanism is suspected to have occurred.
[c] Zebra mussels known to occur on the animal or object.

Source: Reproduced with permission from J.T. Carlton, Dispersal mechanisms of the Zebra Mussel, in T.F. Nalepa and D.W. Schloesser, eds., *Zebra Mussels: Biology, Impacts, and Control* (Boca Raton: Lewis Publishers, 1993), 679.

The invasion of inland lakes in Ontario has most likely been through the introduction of veligers from live-wells and bait buckets, as well as from adult mussels attached to boat hulls or other equipment. While most of the inland sightings have been in waterways that are connected to portions of the Great Lakes, introductions into the Muskoka Lakes and into six eastern Ontario lakes occurred via overland transport. Dispersal to these waterbodies is generally slower than in connected waterbodies. Human-assisted transport is the most likely means by which zebra mussels are introduced to new habitats.

Impact

While such a large and rapid increase in the population of a species is not unprecedented in North America, few if any species have been more visible or have had such immediate ecological and economic impacts as the zebra mussel. These impacts are primarily related to the filter feeding activity of the zebra mussel and their encrustation of solid surfaces.

Zebra mussels are capable of filtering particles ranging in size from bacteria to algae. The removal of phytoplankton from the water column has resulted in increased water clarity. A typical adult zebra mussel, one centimeter in length, can filter more than one litre of lake water per day. On this basis it has been estimated that adult mussels in western Lake Erie are capable of filtering the entire water column several times per day, with a resultant dramatic impact on the concentration of phytoplankton. Reduced levels of phytoplankton in the Great Lakes are not solely related to zebra mussels. A downward trend in phosphorous loading over the past two decades, combined with the impact of grazing by native *Daphnia* species has had a significant impact on phytoplankton. Nevertheless, the decline in phytoplankton abundance and increase in the water clarity throughout much of the Great Lakes region were accelerated after zebra mussels invaded.

One of the greatest concerns about the introduction of zebra mussels to the Great Lakes is the potential impact on native fish communities. Since walleye (*Stizostedion vitreum vitreum*) are light sensitive, increases in water clarity as a result of the filter-feeding activity of zebra mussels could affect their behavior by forcing them into deeper water during the day.[7] Such a pattern has been observed in Lake St. Clair where adult walleye have become more abundant in the dredged shipping channel and less abundant in shallower water since the zebra mussel invasion. Walleye in the St. Lawrence have also shown a preference for deeper water in response to zebra mussel-induced increases in water clarity. Such changes in depth could affect walleye feeding behavior, competitive interactions between walleye and other predators, and the vulnerability of young walleye to predation, but these impacts have yet to be verified.

Another impact of increased water clarity as a result of mussel filtering activity is the proliferation of aquatic macrophytes in areas where lack of light penetration prohibited macrophyte growth prior to the establishment of zebra mussels. The proliferation of aquatic macrophytes following the invasion and build-up of zebra mussels has been documented in several parts of the Great Lakes region.[8] In Lake St. Clair, a shift has occurred from a pelagic system dominated by walleye as the top predator, to a more benthic-oriented system with abundant macrophytes. An increase in the abundance of smallmouth bass, yellow perch, white perch, and northern pike has accompanied these changes, while the abundance of walleye has declined. Parallel changes in fish communities have been observed throughout many parts of the Great Lakes.

Zebra mussels affect the ecology of aquatic ecosystems by causing a shift in energy flow from a pelagic to a benthic food web. Because they devour phytoplankton, the microscopic plants at the base of the

---

[7] B.A. Krishka, *et al.*, *op. cit.*, 44.

[8] R.W. Griffiths, *et al.*, *op. cit.*, 1381-1388.

foodchain, the mussels are competing with algae-eating fish for both food and oxygen. Moderate densities of adult zebra mussel (i.e. 100-1000/m^2) have been estimated to consume 4-18% of the net phytoplankton production, while settled post-veligers probably consume at least an equal quantity of phytoplankton. This may result in a net loss of food resources available for planktivorous fishes and larval fishes, including walleye. As a result, the zebra mussel is regarded as a significant threat to the Great Lakes fisheries, the largest freshwater fisheries in the world.

Benthic communities are also being restructured in areas of high zebra mussel densities, in particular the abundance of benthic invertebrates has been shown to increase. This is because not all of what zebra mussels remove is eaten. What they don't eat is combined with mucus as feces and pseudofeces, and discharged onto the lake bottom where it accumulates. This material provides food for scavenging epibenthic invertebrates while the increased presence of zebra mussel shells provides more substrate for colonizing invertebrates. An increase in the abundance of amphipods and some species of worms and snails has been observed at various locations in the Great Lakes. While this may provide food resources for yellow perch and other walleye forage fishes, it is likely at the expense of pelagic production, which can be important to the walleye food-web. Thus zebra mussels play a substantial role in the nutrient cycling of lakes. They accumulate and store nutrients in amounts similar to those found in emergent and submergent macrophytes or fish populations. In addition they can annually cycle 3-6 times their standing stock of nutrients via their filtering-feeding activity, depositing about 50% of the nitrogen and 40% of the phosphorous directly to the sediment as feces and pseudofeces. In effect, zebra mussel populations act as large nutrient pumps, removing nutrients from the limnetic waters and transferring them to benthic communities.

A further impact of zebra mussels on the benthic invertebrate community is the virtual extirpation of native mussel species. Zebra mussels have been found to colonize unionids to a significantly greater extent than rocky substrates. They are commonly found in and around the gape of unionids, preventing normal valve closure which exposes the unionid to environmental extremes, and preventing valve opening which affects normal metabolic functions such as feeding, growth, respiration, excretion and/or reproduction. Such losses in the diversity of aquatic species can be regarded as indicators of declining environmental health and stability.

A further potential impact on the Great Lakes ecosystem related to zebra mussel filter-feeding is the bio-concentration of contaminants. Only freshwater drum (*Aplodinotus grunniens*) feed heavily on zebra mussels in the Great Lakes, so it is unlikely that fish such as walleye and yellow perch will be affected directly. However the proliferation of amphipods feeding on the feces and pseudofeces of zebra mussels could serve to bio-concentrate contaminants such as polychlorinated biphenyls (PCBs). Amphipods are important food items for most Lake Erie fishes and may become more important with the shift to a benthic food web. Although changes in contaminant burdens of walleye related to this altered trophic structure have not been observed, there are increasing concerns about the cycling of contaminants to top predators such as walleye.

Since zebra mussels represent the introduction of a new food source for wading birds and ducks, resulting in a new link in the food chain, their affect on wildfowl populations and their migration patterns is a growing issue. The potential for waterfowl to accumulate elevated levels of contaminants is a particular concern. Results from trophic transfer experiments conducted in Europe on the tufted duck (*Aytha fuligula*) indicate behavioral disturbances of mature adults, growth retardation, and embryonic mortality as a consequence of pollutant transfer from zebra mussels. It seems logical to assume that some part of the waterfowl community in North America will begin to exploit this new resource. Given the capacity of zebra mussel to accumulate contaminants and its ideal position as a prey item along the banks of water bodies, it would be wise to carefully examine the pathways and flux of organic contaminants in this new part of the ecosystem.

Zebra mussels also have significant economic impacts. These are primarily related to the mussels' tendency to attach themselves to a variety of surfaces, including the in-take pipes of municipal water treatment plants, power plants and industries in the Great Lakes. A square metre of wall at one utility plant was found to contain over 700,000 mussels, while a single in-take pipe at an Ontario water plant was clogged by 30 tons of the mussel. In some cases water flow was reduced by more than 50 percent. Removing the mussels, and preventing further build-up is difficult and expensive. One estimate puts the cost of scraping mussels from pipes in the Great Lakes region alone at $50 to $100 million a year.[9]

Zebra mussels also attach in large numbers to any submerged surface on boats. They cause increased friction and decrease fuel efficiency. Removal of the mussels may also damage or remove paint and surface layers of the boat. The greatest threat to boat owners, however, is related to the possible blockage of internal engine-cooling water passages. Zebra mussel larvae can be drawn into engine passages where they can attach and develop into adult mussels. Small mussel fragments can also be drawn into the engine and damage the water pump impeller. Accumulation of settled zebra mussels on the motor can cause increased friction and damage to moving parts.

Control

No effective technique for the widespread elimination of zebra mussels is known. Control methods tend to use biologically non-specific methods directed at adults (e.g. chlorination) and often require large capital and labor inputs. Problems with these methods include lack of specificity, with consequent danger to other organisms, and the use or production of corrosive chemicals that may damage physical structures. As a result control efforts focus on slowing down the rate of spread of the mussel and easing their effects, with the hope that more effective alternatives for their elimination can be developed over time.

At present, there are six primary methods used to control macrofouling bivalves: these include chlorination, surface coatings, heat treatment, drying, water velocity, and microsieves. In most instances, control of bivalves at a particular water pumping station relies primarily on one of these six methodologies. However the primary method of control is often supplemented with additional control methods specific to the facility, time of year, and mussel life stage. This use of integrated control technology has substantially reduced macrofouling problems caused by bivalves in The Netherlands, where the best solution is a combination of mechanical, physical and chemical applications in declining order of acceptance.

In Canada, such control efforts tend to concentrate on specific problems related to zebra mussel infestation, such as blocked water-intake pipes and damage to the cooling systems of boats. Ontario Hydro is the lead organization involved in research into the control of zebra mussels at intake pipes. Chemical treatments such as chlorine have been used in the past; however, high chlorine levels can produce carcinogenic organic side effects with severe environmental repercussions. Heat treatment is also being experimented with in a limited context (e.g. certain parts of nuclear power plants) but this approach also has deleterious impacts on the aquatic environment.

At the regional level, the Ontario Ministry of Natural Resources (MNR) organized two major initiatives in 1992 to slow down and monitor the spread of zebra mussels. The first was a series of public education initiatives aimed at getting public assistance and support in the fight against the spread of mussels. The purpose of the program was to provide current information about zebra mussels. Brochures, posters,

[9] P.D.N. Herbert, C.C. Wilson, M.H. Murdoc and R. Lazar, Demography and ecological impacts of the invading mollusc *Dreissena Polymorpha, Canadian Journal of Zoology*, 69, 1991, 405-409.

outdoor signs, displays, videos and slide shows have been produced, in addition to a "hotline" number[10] and an annual newsletter – the Zebra Mussel Watch. The latter was to provide information on the spread of mussels, government initiatives to control the problem and useful tips and advice for the general public about how to slow down the spread of the mussel. Much of this information was directed at boat owners who were encouraged to inspect and remove zebra mussels from boats and motors as a way of slowing the spread of the mussels and reducing their boat repair costs.[11] Only one issue of the newsletter was published as a result of cutbacks in government spending.

The second major initiative launched by MNR was the Invasion Monitoring Program, which was to monitor the distribution, spread and abundance of zebra mussels over time. More than 150 monitoring stations were established throughout the Great Lakes and major waterways such as the Trent Severn and Rideau Canal System. The presence of veligers in different lakes and rivers throughout the province was tested by plankton pumping. Steel plates were placed in different water bodies to monitor the presence of settling vilegers over time, while plates and cement blocks were used to detect the presence of adult zebra mussels. These stations were to provide vital information about the spread of zebra mussels. Again, however, the program was discontinued because of a lack of government funding.

Two alternative approaches to the control of zebra mussels include the use of biological controls and the targeting of zebra mussels by interfering with their unique physiology. Biological controls include diving ducks and some fish that feed on zebra mussels. The freshwater drum (*Aplodinotus grunniens*) is the most likely fish predator since it is the only fish with pharyngeal teeth capable of crushing mollusk shells. Studies of predation on zebra mussel by freshwater drum have found that predation increased as drum size increased; freshwater drum over 375 mm feed heavily on zebra mussel and may become a possible biological control mechanism for mussels in portions of North America.[12] Predation by waterfowl would, however, be limited to the warmer months when the Great Lakes are not frozen over and could result in reproductive problems due to bioaccumulation of contaminants. Thus predators are unlikely to become an important control.

Rather than attempting to control adult zebra mussels, intervening at an earlier and more sensitive stage in the life cycle may prove to be more effective. As the larva develops, its tolerance limits for various environmental conditions increase. For example, larvae can tolerate a temperature range of between 0° and 30° C. Temperatures outside this range can be used to kill off the larvae before they infest a location (e.g. nuclear power plants). If a method can be developed that disrupts the normal reproductive cycle of the zebra mussel, then population densities may be reduced.

Research

Assessing the potential impacts of the zebra mussel on native biotic communities and developing successful control strategies requires a thorough understanding of its basic life history patterns in its new environment. The eco-physiology (i.e. the study of how an organism is adapted to its environment) of freshwater bivalves is poorly understood when compared with economically important marine species. The appearance of the zebra mussel thus offers an opportunity to study the adaptation of an organism to a new freshwater environment. In order to be successful in its new habitat, the organism must be able to

[10] The Ontario Federation of Anglers and Hunters joined with the Canadian Coast Guard and the Ontario Ministry of Natural Resources to set up a 1-800- Hotline. Boaters, snorkelers and divers are asked to report sightings of exotic aquatic plants and animals. The hotline number is 1-800-563-7711.

[11] Ministry of Natural Resources, Zebra Mussels: A Guide for Boaters and Cottagers, Queen's Printer for Ontario, 1995.

[12] J.R.P. French and M.T. Bur, Predation of the Zebra Mussel (*Dreissena Polymorpha*) by Freshwater Drum in Western Lake Erie, in T.F. Nalepa and D.W. Schloesser, *op. cit.*, 453-464.

adapt to appropriate seasonal cues regulating important life history events, such as reproduction. Because the ecology of the Great Lakes has been studied in detail, the impact of the zebra mussel on other components of the ecosystem can more readily be isolated and examined.

It is increasingly recognized that the best approach to controlling zebra mussels may be to target critical physiological systems in the zebra mussel, e.g. the nervous system and the cardiovascular system. Research into the anatomy and physiology of the zebra mussel nervous system may provide a means of targeting zebra mussels, since it controls such functions as feeding, reproduction, and its main defense tactic – valve closure. To evaluate whether this system can be targeted, research efforts are currently focusing on how the system functions normally, especially on finding out what transmitters are being used as messengers. Research into internal biochemical regulators of reproduction suggests that serotonin, a neurotransmitter in mussels and mammals, can activate spawning in both male and female zebra mussels.[13] The identification of serotonin as the internal physiological activator of spawning has also led to experiments directed at cloning the gene mediating the spawning response. Such research may be valuable not only in predicting its further spread, but also in investigating basic mechanisms of reproduction and development and in developing new strategies to mitigate its impact. Notwithstanding the negative ecological and economic impacts of this exotic species, the prolific reproductive capabilities of the zebra mussel are regarded as a vital resource for studies of reproductive mechanisms and capabilities.

Research into external environmental cues focuses on stimuli from outside an organism that activate behaviors. For example, water temperature has been implicated as the primary factor regulating gonadal maturation and spawning. Temperatures above 12°C have been reported as necessary for spawning to begin, enabling the synchronization of seasonal maturity processes so that mass concentrations of eggs and sperms will be released.[14] In heated lakes, spawning begins earlier and persists later into the fall than in nearby non-heated lakes. However, increased water temperature alone may be insufficient to trigger spawning. Increasing phytoplankton abundance has been identified as an important external chemical cue triggering spawning in marine invertebrates and is likely to cause variability in zebra mussel spawning and veliger abundance. Another important factor influencing distributions of freshwater mollusks is pH; a decrease in pH is known to have a negative impact on the abundance of bivalves.

Because the zebra mussel is believed to be a relatively recent "immigrant" to freshwater, as evidenced by the retention of the veliger stage of marine forms, zebra mussels have a higher resistance to salinity when compared to freshwater unionids. The zebra mussel is widespread in estuaries and inland brackish waters of Europe. Within these habitats, its distribution and abundance are limited by salinity, availability of hard substratum ice scour, exposure and perhaps turbidity. High salinity is the preeminent limiting factor; upper tolerance limit ranges from about 0.6 to 12% salinity. Analysis of the European literature suggests that zebra mussel will become widespread and abundant in estuaries and brackish waters in North America.

Like many other external factors, the substrate represents a limiting factor that can be manipulated in the control of zebra mussels. It is known that the occurrence of zebra mussels is limited to localities with solid structures. Lake bottoms in The Netherlands, which consist of sand, silt, clay and a mixture of these substrates, are thus unsuitable for zebra mussels and zebra mussels colonies are restricted to submerged vegetation, stones, and shells of dead and live mussels. Just as in The Netherlands, where the addition of a suitable substrate is being used to increase zebra mussel numbers, the removal of suitable substrate can reduce their numbers. Studies have also been conducted to determine if zebra mussel have a preference

[13] J.L Ram, P.P. Fong and D.W. Garton, *American Zoologist*, 36, 1996, 326-338.

[14] M. Sprung, Field and laboratory observations of *Dreissena Polymorpha* Larvae: abundance, growth, mortality and food demands, *Arch. Hydrobiol.*, 115, 1989, 537-561.

for different construction materials.[15] The following preferences were exhibited – copper, galvanized iron, aluminum, acrylic at the low end and pine, polypropylene, asbestos and stainless steel at the upper end. Thus, using materials containing aluminum, iron and copper can reduce colonization of zebra mussel over a short term.

Future

Given the European experience of the past two centuries, the arrival of zebra mussels in the Great Lakes and inland waterways of North America is probably permanent. The common pattern of zebra mussel invasions in Europe is a phase of rapid population growth, followed by sudden decreases to low steady-state population densities. The colonization of the Great Lakes by zebra mussels is still in its initial stage, and it is too early to anticipate the potential impact they will have on this aquatic ecosystem. Continued study is necessary in order to determine relationships between environmental factors regulating reproduction and long-term mussel population dynamics.

It is possible that not all impacts will be judged negative in the longer term. While the ecological impacts of zebra mussels are almost invariably described as detrimental to the Great Lakes region, applied research in Europe has concentrated increasingly, over the past decade, on the positive applications of the zebra mussel for water management and its role in ecosystems.[16] For example, the high filtration activity of the mussel has successfully reduced the high abundance of phytoplankton in many parts of continental Europe. It is also the main source of food for diving ducks which winter in large numbers in The Netherlands and for benthivorous fish such as roach and carp. As a result of these benefits, management authorities in Lake Volkerakmeer, The Netherlands, have artificially increased the density of zebra mussels by adding marine shells as a substrate for attachment and are using zebra mussels to increase water transparency. In the last decade, mussels have effectively concentrated polluted suspended matter by transforming the very small, suspended particles into fecal pellets with much higher sinking rates. Large numbers of macrophytes and waterfowl now inhabit the lake.

In the absence of a clear understanding of community interactions and the factors that control them, a cautious management approach is recommended. Initial responses to the introduction of zebra mussels from government agencies charged with responsibility for natural resource management was encouraging. However, some complacency combined with substantial cutbacks in government spending seems to have resulted in a scaling down of programs designed to understand, monitor and control the zebra mussel problem. On-going government funding and commitment is essential to the maintenance of monitoring and control efforts, including public training and education regarding zebra mussels and other introduced species.

These efforts become increasing important considering the evidence that several recently introduced species had origins in the Caspian Sea/Black Sea region, and that potentially other *Dreissena* varieties may also have been transported to North America. At least two more varieties of *Dreissena polymorpha* and some varieties of *Dreissena rostriformis* are native to this region. The latter species is able to tolerate more brackish conditions than *Dreissena polymorpha* and, if introduced to North America (if not already present) may colonize areas not suited for *Dreissena polymorpha.* The quagga mussel (*Dreissena bugenis*), a close relative of the zebra mussel, was introduced to the Great Lakes from Europe in the 1980s. The quagga mussel is similar in appearance and biology to the zebra mussel, and has similar

[15]B.W. Kilgour and G.L. Mackie, 1993. Colonisation of different construction materials by the zebra mussel (*Dreissena polymorpha*), in T.F. Nalepa and D.W. Schloesser, eds., *op. cit.,* 167-173.

[16] H. Smit, A. bij de Vaate, H.H. Reeders, E.H. van Nes, and Noordhuis, Colonization, ecology, and positive aspects of zebra mussels (*Dreissena polymorpha*) in The Netherlands, in T. F. Nalepa and D. W. Schloesser, eds., *op. cit.,* 55-77.

effects on the ecosystem but can live in deeper and colder water. Future research should compare genetic and morphological characteristics of Dreissena species in North America and in the Caspian Sea/Black Sea region in order to better predict the future range of Dreissena infestations in North America.

Questions

1. Because the adverse effects of toxicants on the zebra mussel may affect the aquatic food chain, the zebra mussel represents a valuable bio-monitoring organism and an early warning system. Explain, using examples, the importance of this.

2. From a research perspective, the zebra mussel has several positive attributes; it plays an important role in the aquatic food chain, is easy to collect and handle, and has low control mortality. Outline some potential research experiments using the zebra mussel as a test organism.

3. For divers, snorkelers, tourists, and cottage owners, the increased water clarity associated with zebra mussel infestation is a bonus. To what extent could this perceived benefit impact on the management of the species?

4. Compare the biology, impact and control of the zebra mussel with that of another exotic species. Explain the similarities and differences with respect to the implications for the management of both species.

Further Readings

Anon. 1996. *The Biology, Ecology, and Physiology of Zebra Mussels.* Special Issue, *American Zoologist.* June.

Herbert, P.D.N., Wilson, C.C., Murdoc, M.H. and R. Lazar. 1991. Demography and ecological impacts of the invading *mollusc Dreissena polymorpha. Canadian Journal of Zoology,* 69: 405-409.

Krishka, B. A., Cholmondeley, R.F., Dextrase, A.J. and P.J. Colby. 1996. Impacts of introductions and removals on Ontario percid communities. Percid Community Synthesis, Introductions and Removals Working Group. Peterborough: Ontario Ministry of Natural Resources.

Nalepa, T.F. and D.W. Schloesser (eds.) 1993. *Zebra Mussels: Biology, Impacts, and Control.* Boca Raton: Lewis Publishers.

Web Sites

- Zebra Mussels: http://octopus.gma.org/surfing/human/zebra.html
- Zebra Mussel Clearinghouse: http://nfrcg.gov/zebra.mussels
- Zebra Mussels, Great Lakes: http://www.noaa.gov/public-affairs/pr96/mar96/noaa96-11.html
- Zebra Mussels, Michigan: http://www.boatersinternational.com/dnr/harbors/MUSSELS.html
- Zebra Mussels, South Burlington: http://www.gmpvt.com/pressrel/pr960315.htm
- Zebra Mussels, Wisconsin: http://kapalua.mhpcc.edu/waterWatch/zebra/html/wi.shtml

Audio-Visual Material

- *The Invaders*, Canadian Broadcasting Corporation, 1993 (47 min.).

Case Four
Temagami Old-Growth Forest: Experiments in Community-Based Management

Focus Concept

The potential of community-based management to resolve complex land and resource management disputes.

Introduction

In 1988, the Provincial Government of Ontario announced, without public consultation, the construction of logging roads into the old-growth red and white pine forests of the Temagami region, northeastern Ontario. Clear-cutting of the world's second largest stand of old-growth red and white pine was to be allowed. This was to take place in the heartland of the Teme-Augama Anishinabai, an area of unresolved Native land claims. The Teme-Augama Anishinabai unsuccessfully sought a court injunction to stop the plan. With other concerned people, mostly environmentalists, they mounted road blockades on the controversial Red Squirrel forest access road, which resulted in over two hundred arrests. The protests gained media attention, captured the interest of the public, and drew responses from the provincial government. The anti-logging alliance eventually succeeded in stopping road construction.

As a result of the controversy, the government of Ontario adopted a variety of approaches in an attempt to resolve resource use conflicts in the area. A key aspect of these approaches has been the decentralization of responsibilities to local communities. The present case study addresses three of these initiatives.

The first is the Temagami Area Comprehensive Planning Program (CPP), launched in 1989 at the behest of the government, to establish "model" management of the resources of the Temagami region. A Comprehensive Planning Council (CPC) was established as an advisory body to the program, comprised of local representatives who managed an extensive six-year public consultation process. The CPC made its recommendations for a comprehensive land-use plan to the Minister of Natural Resources (MNR) in April 1996. These recommendations provided the basis for the Temagami Land-Use Strategy, released to the public two months later by the Government of Ontario. The Strategy adopted most of the CPC's recommendations, including a preference for the protection of old-growth "representative sites" rather than a moratorium on old-growth harvesting. This land-use decision put Temagami back in the national spotlight. In the fall of 1996, in scenes reminiscent of the Red Squirrel incident, environmentalists attempted to block the entrance to the Owain Lake forest. Over forty arrests were made.

The second initiative stems from attempts to resolve the conflict between the Government of Ontario and the Teme-Augama Anishinabai, the indigenous people of the region. The two parties signed an agreement in 1990 to negotiate a Treaty of Co-existence, and to share stewardship responsibility for land and resource management in part of the Temagami region. The Wendaban Stewardship Authority (WSA) was established in May 1991 with equal representation of Teme-Augama Anishinabai-appointed Natives and government-appointed non-Natives. Hailed locally as "a new approach to cooperative management" and "a working model for co-existence with aboriginal people,"[1] the Authority succeeded in achieving consensus among diverse interests over resource management across four townships in the heart of the

[1] A. Hakala, Stewardship hailed as new approach, *The North Bay Nugget*, November 23, 1992, A4.

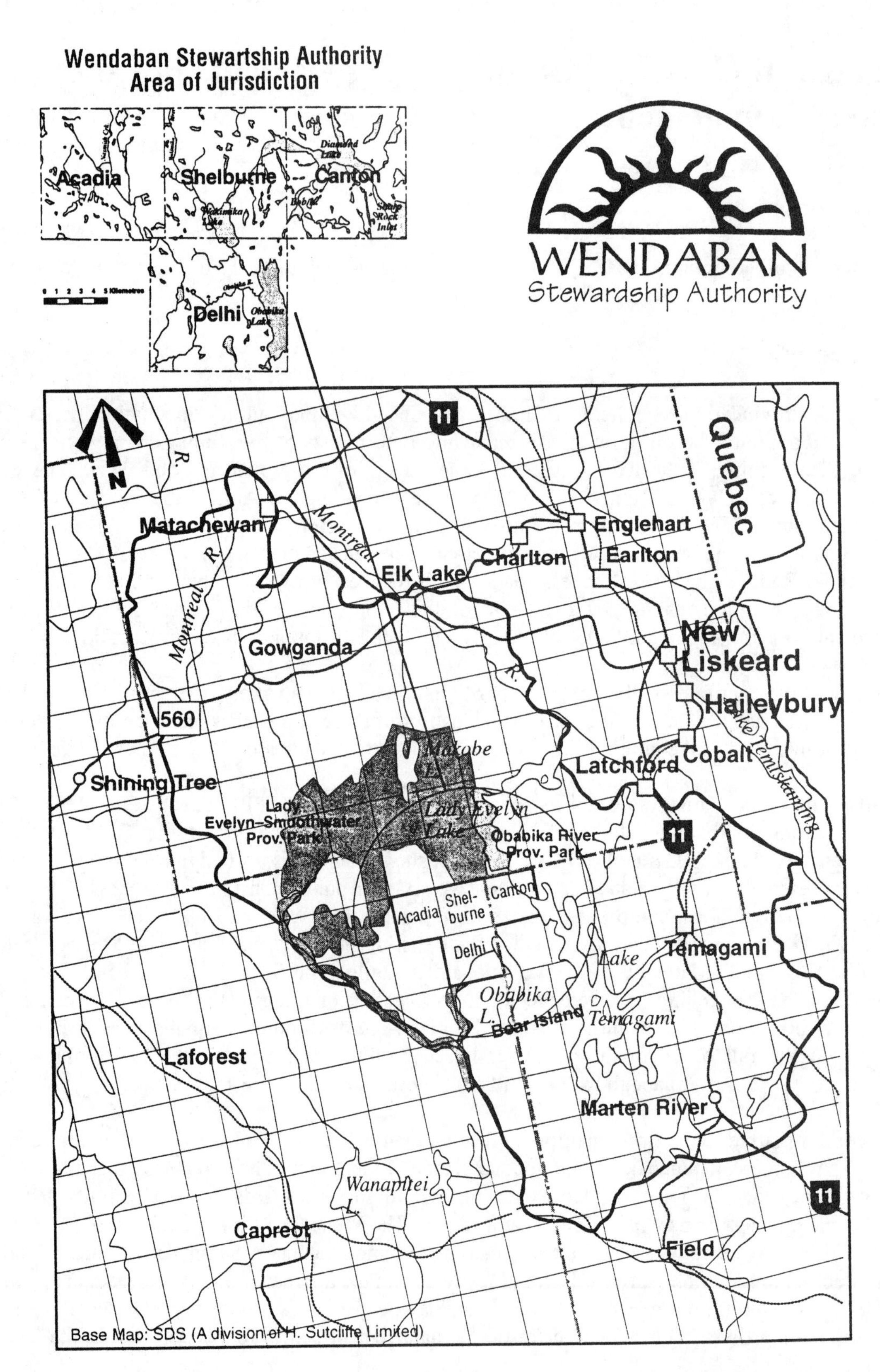

Figure 4.1. Temagami Region, including Area of Jurisdiction of WSA

traditional homeland of the Teme-Augama Anishinabai. Despite its achievements, the future of the WSA is currently in doubt.

A third initiative is the Elk Lake Community Forest Project (ELCF), established under the provincial government's Sustainable Forestry Program. The ELCF promotes local input on issues of balancing the priorities of development and forest conservation at Elk Lake, a northern community of some 550 people who depend on logging for their livelihoods.

This case study outlines the structures and processes involved in these initiatives, and discusses possible reasons for their successes and limitations.

Old-Growth Forests

The natural range of old-growth white pine forests once extended over much of the eastern United States and southeastern Canada to cover an estimated area of 6 Mha. Less than one percent (24,000 ha) of this ecosystem remains today, which classifies this ecosystem as "endangered".[2] Comparable estimates are unavailable for old-growth red pine but it has also been severely over exploited and is considered likely to be endangered.[3]

In Canada, the present distribution of old-growth red and white pine is limited to a relatively narrow band that runs across central Ontario, from the Temagami area in the east to the shores of Lake Superior, as well as some parts of northwestern Ontario. Almost 4 Mha of forest land in Ontario contains a component of red and white pine; concentrations range from 10 percent to more than 40 percent on any given stand. Of these areas, more than 124,000 ha are classified as old-growth pine and 66,000 ha are protected. Old-growth forests are defined by the Ontario Ministry of Natural Resources (MNR) as forests older than 120 years that show little or no traces of activities such as logging. Nearly 11,000 ha of the province's 66,000 ha of protected old-growth pine forest are in the Temagami area. Two major stands are present; the Lake Obabika North stand and Owain Lake stand.

The Lake Obabika North old-growth site supports Temagami's highest concentration of old-growth red and white pine forest and is the second largest old-growth red and white pine forest ecosystem in North America.[4] This is also the site that was the focus of the controversy over Temagami's old-growth forests in the late 1980s. The site is approximately 2,400 ha in size and contains trees that are 130 to 300 years old, with a maximum height of 30 m. The area shows no signs of logging.

The Owain Lake stand is located about 45 km southeast of the town of Temagami, and 60 km from the Lake Obabika old-growth site. It is composed of approximately 1,400 ha and is regarded as the third largest of the known old-growth white pine-dominated stands in North America.[5] The ages of the trees range from 75 to 145 years, with a mean of 117. There is some evidence that harvesting may have occurred at the turn of the century and again in the late 1930s or early 1940s.

The Ontario MNR has adopted a conservation strategy for old-growth red and white pine ecosystems in which it takes "a two pronged approach" to their protection and management; some sites are set aside for

---

[2] P.A. Quincy, A Critique of the Proposed Management of Old-Growth White and Red Pine Forest in Temagami, Ontario Resulting from the Comprehensive Planning Process of 1996 with a Case Study Analysis of the Owain Lake Old-Growth Pine Stand as a Representative Ecosystem. Submission prepared for Earthroots, http://www.web/net/~eroots/erquinresp.html, 1996:1.

[3] *Op cit.*

[4] Bark Lake Conservation Reserve is the largest old-growth red and white pine forest ecosystem in North America.

[5] *Op. cit.*, 4.

protection, while others are available for timber harvesting. The strategy gives a commitment to the protection of representative sites of old-growth pine forests so that they can evolve through natural ecological processes. The expansion of the boundaries of Obabika River Provincial Park by 3,520 ha in 1996 is an example of the application of this strategy since it resulted in the protection of the Lake Obabika North old-growth site. A range of recreation- and tourism-oriented activities are permitted within the old-growth site while commercial timber harvesting, mining and mineral exploration, hydroelectric development, new road construction and other industrial activities are prohibited.

In addition to protection and regeneration, the conservation strategy provides for "a sustainable supply of red and white pine to the forest industry." Sites such as the Owain Lake stand which are younger than 120 years and show signs of selective logging do not qualify for protection as representative old-growth sites. As a result, the Temagami Land-Use Strategy has created a harvest block of 321 ha on the Owain Lake site in which partial logging is permitted on 228 ha while the remaining 93 ha are a no-cut reserve. Species to be harvested include jack pine, spruce, poplar and white birch, as well as red and white pine. Clear-cutting is prohibited under the land-use strategy, and harvesting must be carried out using the "shelterwood system". This method is reputed to encourage natural regeneration by leaving healthy mature trees standing along with standing dead trees and dying and rotting trees. Tree tops and limbs of harvested trees are to be left on the forest floor to provide wildlife habitat. Logging of the Owain Lake site commenced in the fall of 1996, triggering the latest series of protests and blockades by environmental activists.[6]

The Temagami Area Comprehensive Planning Program (CPP)

The Temagami Area Comprehensive Planning Program (CPP) was launched in July 1989 by the Minister of Natural Resources in an effort to establish "model" management of the resources of the Temagami area. The program followed the acceptance by provincial government of several recommendations made by the Temagami Area Working Group (TAWG), a citizens' advisory committee to the Minister struck in 1987 to address land and resource conflicts in the area. A second citizen's advisory committee - the Temagami Advisory Council (TAC) – was established to recommend a comprehensive plan to the Minister by March 31, 1992.

The TAC was replaced by the Comprehensive Planning Council (CPC) in May 1991 as part of an attempt by the Ministry of Natural Resources to "... strengthen the role of local communities in the management of natural resources in the Temagami area." Membership on the Council, which included teachers, local business people, and residents, was expanded from nine to thirteen people, and the deadline for submission of the comprehensive plan was extended by two years. A comprehensive Planning Team was established to support the CPC in its work. Terms of reference for the CPC included the following:

- Recommend to the Minister of Natural resources, in an advisory capacity, a Comprehensive Plan for the area, excluding the area managed by the WSA (see below);
- Manage the public consultation process in the development of the Comprehensive Plan;
- Provide advice on on-going land use and resource management decisions, until such time as the Plan is approved;
- Provide advice to the Minister of Natural Resources regarding alternative mechanisms for third party interests to provide input into the negotiations with the Teme-Augama Anishinabai.

In September 1992, the CPC launched a year-long public consultation process on behalf of the province and the Teme-Augama Anishinabai as part of the negotiations for a Treaty of Co-existence. Five representatives of the Teme-Augama Anishinabai were appointed to the CPC during the course of these

[6] See the Earthroots web site for more details: http://www.web.net/~eroots

discussions. The deadline for completion of the Comprehensive Plan was again extended, to March 31, 1996.

After September 1993, the CPC turned its attention to discussions about various land-use options for resource management in the Temagami area (excluding the four townships managed by the WSA). Three alternative land-use scenarios were presented to the public in May 1994. These reflected a range of potential land-use alternatives, degrees of commercial development, and resource protection. Open houses, workshops, and various meetings were held over the summer of 1994 to solicit public reaction to the plans. A comprehensive land use map was circulated a year later, in June 1995, reflecting community feedback on how to accommodate the many competing resource demands in the region.[7] CPC released a draft land-use proposal at the end of 1995, followed by several issue-resolution sessions in early 1996 with representatives of local, regional and provincial organizations. The Teme-Augama Anishinabai declined to contribute to these sessions.

The final recommendations of the CPC were submitted to the Natural Resources Minister in April 1996 after more than six years of public debate and consultation. The recommendations fall into two categories: land use recommendations in relation to the area's Crown land recreation, cultural heritage, fisheries, fire, landscape ecology, minerals, provincial parks, tourism, timber and wildlife values; and other recommendations which address issues such as governance, the interests of and need to involve the area's aboriginal peoples, future resource management planning, and education.

On June 28, 1996, the Minister released the Comprehensive Land-Use Strategy for the Temagami Area to the public. According to MNR, the strategy adopted twenty-two of the CPC's thirty-nine recommendations without change and the remaining seventeen with minor changes. For example, MNR adopted CPC's recommendation for the development of resource plans for forestry, fisheries, parks and other resources in the area, and agreed to develop a forest management plan to cover 1997-2002 and a park plan for release in spring 1998.[8] MNR also adopted the CPC recommendation that a total of 12 representative old-growth sites (44 percent of the old-growth pine area in Temagami) be protected in the area. In fact, MNR claims to have increased the proposed protection level from 44 percent to 52 percent, with a longer-term goal of achieving a protection level of 65 percent at both the provincial level and within the Temagami area.[9] CPC supported MNR's representative approach to the protection of the old-growth forests in preference to a moratorium on harvesting. This included CPC approval of MNR's decision to protect the North Obabika Lake site while permitting partial logging of Owain Lake (see above). At the same time, MNR rejected the CPC recommendation to protect the headwater areas of the Lady Evelyn-Smoothwater Provincial Park. The Land-Use Strategy allows forestry and mining, "subject to special conditions protecting headwaters values" in the headwater area to the east of the park.[10]

The land-use strategy establishes four new land-use zones, in addition to the seven provincial parks already in place, as the basis for all future land-use planning in the area. These zones are designed "to ensure that the interests of all resource users – both recreational and commercial – can be accommodated, while ensuring that significant ecological and other natural resource values are protected and managed."[11] These land-use zones were recommended by the CPC. They are:

---

[7] At the direction of the provincial government, this included the Wendaban Stewardship Authority's Forest Stewardship Plan, submitted in February 1995; see section below.

[8] Ministry of Natural Resources, Response of the Government of Ontario to the Temagami Comprehensive Planning Council recommendations (http://www.mnr.gov.on.ca/MNR/temagami/temares.html) 1996, recommendation # 2.

[9] *Op. cit.*, recommendation #26.

[10] *Op. cit.*, recommendation #24.

[11] Ministry of Natural Resources, Description of Land-Use Zones in the Temagami Comprehensive Planning Area, Fact Sheet, June 1996.

- Protected Areas where commercial activities such as timber harvesting and mining are prohibited;
- Special Management Areas where resource extraction, as well as existing and potential recreation and tourism opportunities, are permitted under special prescriptions to ensure protection of significant resources;
- Integrated Management Areas where resource extraction and development and recreational activities will be integrated, allowing commercial timber harvesting, mining and aggregate extraction to take place alongside recreational uses; and
- Developed Areas where existing development is recognized and additional development is permitted, while ensuring it is compatible with other uses.

The land-use strategy has been described by the provincial government as a planning tool designed "to enhance environmental protection in Temagami." In the words of Natural Resources Minister Chris Hodgson:

> This land-use strategy sets up a framework that will protect the environment, improve community stability and after years of uncertainty, get on with the job of enhancing the quality of life in the Temagami area.[12]

The environmentalists are less than happy with the strategy and have been quick to point to departures by MNR from the recommendations of the CPC. Tim Gray, executive director of Wildlands League, described the failure to adopt CPC's recommendation for the protection of the headwater areas of the Lady Evelyn-Smoothwater Park in the following terms:

> Derailment of the CPC's land use plan throws the region wide open to ecological devastation, as well as wasting millions of dollars of public funds and years of consultation.[13]

In the same statement, Gray describes the CPC's recommendations as "the last chance to protect wilderness and old-growth forests throughout the Temagami region" and "a reasonable compromise to conserve wilderness and permit development in Temagami." Yet only six months earlier Gray was part of a coalition of five environmental groups that put forward an alternative land use plan to the CPC process. Gray provided the following explanation for their proposal:

> We've followed the Comprehensive Planning Council process closely over the years. The key difference between the CPC plan and what we're proposing is that our proposal resolves the land use conflicts and keeps the benefits of Temagami's spectacular natural values in the community. Unfortunately, the current CPC plan would just perpetuate the status quo, meaning a continued loss of natural values, and a continuous flow of the benefits of industrial activities out of the region.[14]

The proposed plan which was described as "a balancing act between environmental protection and industrial activities"[15], included the establishment of a number of reserves and a new "forestry authority" which would support local control over mining, logging and tourism activities in the region.

---

[12] Chris Hodgson, Minister of Natural Resources as quoted in Ministry of Natural Resources, Province adopts strategy to enhance environmental protection in Temagami, News Release, June 28, 1996.

[13] Tim Gray, executive director, Wildlands League, as quoted in http://web.idirect.com/~wildland/temagami.html, 1.

[14] Wildlands League, Earthroots, Northwatch, Friends of Temagami, and TEAC, New and Balanced Land Use Plan for Temagami, Press Release, 31 January 1996, 2.

[15] *Op cit.*, 1.

This raises a number of interesting questions: to what extent should government be obliged to adopt the recommendations of local people gathered at considerable expense over a number of years of consultation? Is it possible or desirable for government to attempt to appease all interest groups in a land use planning exercise? If not, to whose interests should government cater? To what extent should environmental activists interfere in decisions relating to local land and resource use?

Wendaban Stewardship Authority (WSA)

The Wendaban Stewardship Authority was established in May 1991, under an Addendum to the Memorandum of Understanding (MOU), signed in April 1990 between the Government of Ontario and the Teme-Augama Anishinabai. The MOU also set up negotiations for a treaty covering 10,000 km^2 of land. This land – known as n'Daki Menan - has been the centre of a long-standing dispute between the Teme-Augama Anishinabai and the Ontario government. Because the lands of the Teme-Augama Anishinabai were not included in the 1850 Robinson-Huron Treaty, and their leadership never signed the treaty, the Teme-Augama Anishinabai claim that they never surrendered title to their land. An official reserve, one square mile in size, was established in 1971 on Bear Island in Lake Temagami, essentially as a measure to provide space for village infrastructure. However, the Teme-Augama Anishinabai continued to claim a much larger area comprising 10,000 km^2 over 110 townships within the Temagami area.

In 1973, as part of their intention to advance a claim to ownership of their traditional territory, the Teme-Augama Anishinabai filed formal land cautions with the Registrar of Titles on the claim area. The land caution did not affect commercial forestry operations, but prevented the staking of new mining claims and new tourism-related economic development.

Efforts by the Teme-Augama Anishinabai to assert ownership of their lands culminated in a 1991 Supreme Court of Canada decision that the Teme-Augama Anishinabai adhered to the 1850 Robinson-Huron Treaty without actually signing it. Aboriginal title to their land had been extinguished, according to the Supreme Court, when some members of the band accepted 4 dollar treaty payments and because Bear Island had been made into a reserve in 1971. The Court also said that the Crown had breached its fiduciary obligations to the Teme-Augama Anishinabai and acknowledged that "these matters currently form the subject of negotiations between the parties." [16]

In addition to treaty negotiations, the MOU included a bilateral process that guaranteed the Teme-Augama Anishinabai an advisory role in the Ontario Ministry of Natural Resources (MNR) timber management planning process for the Temagami district. There was a commitment from both parties to establish a 'stewardship council' for part of the area, an arrangement that was hailed as a solution to the dispute.[17] An addendum to the MOU, signed in May 1991, brought the council into existence as the Wendaban Stewardship Authority (WSA).[18] The authority, which has been called Canada's first joint provincial-native governing body, was given jurisdiction over four townships (roughly 400 km^2) northwest of Temagami and within the traditional homeland of the Teme-Augama Anishinabai. While most of the area is Crown land, it includes a few patented mining claims and privately owned cottage lands. The Wendaban stewardship area also includes the old-growth white pine forest at the centre of the environmentalist and native blockades in the late 1980s, when the controversial Red Squirrel forest extension was being built to allow loggers access to old-growth forest.

---

[16] Land cautions in Temagami, in place since being filed by the Bear Island First Nation in 1973, were lifted by the Ontario courts in November 1995. Previously, the cautions had shielded the 133 townships from development.
[17] Royal Commission on Aboriginal Peoples, Report of the Royal Commission on Aboriginal Peoples, Vol. 2, Part II, Appendix 4b (Ottawa, 1996), 754.
[18] "The authority was named for Wendaban, who was head of the principal Aboriginal family that traditionally occupied the stewardship area. Wendaban means 'whence the dawn comes'"; *op. cit.*, 754.

The authority was set up as a decision-making body to report simultaneously to the provincial government and the Teme-Augama Anishinabai. In this regard, it represented the promise of shared jurisdiction within the stewardship area, not mere advisory powers. According to its terms of reference, the WSA is responsible for monitoring, undertaking studies of and planning for all uses and activities, ranging from recreation and tourism, fish and wildlife to land development and cultural heritage – responsibilities that had previously been under the control of the local MNR office.

The authority is comprised of equal representation from the two parties; the Teme-Augama Anishinabai, and the province of Ontario appoint six members each. Members are selected with a view to incorporating the diversity of local interests in the planning and management process. None of the members is a provincial public servant. Ontario's representatives included a local township reeve, the manager of a nearby sawmill and a local environmental activist; the Aboriginal representatives included the owner of a contracting business, a trade unionist and the manager of a craft co-operative.[19]

A non-voting chair is appointed by mutual agreement, with some aspects of traditional Aboriginal protocol, such as consensus-based decision-making, built into procedure.

The WSA advocates holistic land and resource use based on four principles: sustained life, sustainable development, coexistence between Aboriginal and non-Aboriginal peoples, and public involvement in the activities of the authority. These principles are reflected in the Comprehensive Forest Stewardship Plan put forward by the WSA in 1993. In developing the Plan, the WSA conducted a series of meetings with interested groups and individuals to establish a process for public participation, to identify issues, and to hear public response to the goals and objectives of the Plans. In October of 1993, the WSA hosted a final public review and critique of the planning options and zoning decisions in its Forest Stewardship Plan.[20] The plan covers all aspects and uses of the forest, including access, timber, fish and wildlife, cultural heritage, recreation and tourism. Zoning for various land uses is a central framework for the plan. Zoned activities include: designation of areas for protection, mining, timber extraction and recreation-tourism-wilderness enjoyment.

The WSA, despite notable successes, now faces an impasse that puts its future viability in question. The achievements and the limitations of the model were reviewed by the Royal Commission on Aboriginal Peoples. In regard to its achievements, the WSA was found to have worked much better at achieving consensus among divergent interests than many had anticipated:

> There were initial doubts about the potential for success using a consensus approach to decision making, given the diverse cultures and backgrounds of the members, the previous level of conflict over resource management issues, and the fifty-fifty split in representation. These doubts seemed to be backed up by conventional wisdom on resolving conflict: consensus is suited for situations where the level of conflict is low and groups have much in common. Once the members and chair formed a comfortable rapport, however, the authority established an informal routine for decision making.
>
> In fact, in more than three years of operation, not one of the authority's decisions split the membership on Aboriginal/non-Aboriginal lines. The main points of tension were between those in favour of and those opposed to resource development, and there were Aboriginal and non-Aboriginal members in both camps. In June 1994, members reached

[19] *Op. cit.*, 755.

[20] Ministry of Natural Resources, Chronology of public consultation on Temagami land use planning, Fact Sheet (http://www.mnr.gov.on.ca/MNR/temagami/temachro.html), 1996.

> consensus on a 20-year forest stewardship plan for the area under the authority's jurisdiction, a plan that establishes land use zones and regulates all uses, including recreation, timber, mining, wildlife, water and cultural heritage. That plan was subsequently submitted to Ontario and the Teme-Augama Anishinabai.[21]

Seriously undermining the functionality of the WSA was the failure of the Ontario Government to keep up its end of the bargain, which reinforced the tendency of district MNR staff to resist the sharing of jurisdiction in *practice*:[22]

> While the Teme-Augama Anishinabai sanctioned WSA as a decision-making body through a general assembly resolution, the authority did not obtain the promised legislative jurisdiction over the four townships from Ontario. The jurisdictional vacuum was temporarily alleviated through the tacit agreement by all parties to act as if the authority possessed the appropriate legislative base.[23]
>
> While useful, this strategy caused some difficulty in practice. On more than one occasion, authority decisions that were contrary to policies of the local planning board were challenged by district staff of the ministry of natural resources (sic) because of WSA's lack of legislated decision-making authority. Although the situation was resolved by the minister directing his staff to acknowledge the authority's jurisdiction, it illustrates the problem of operating without a clear legislative base.[24]

With the election in June of 1995 of a new Conservative government to replace the New Democratic Party, right-wing disapproval of power sharing with aboriginal and grassroots interests displaced the social democratic climate in which the WSA, and its broader policy context, had developed. An agreement-in-principle on a Treaty of Coexistence, reached between Ontario and the Teme-Augama Anishinabai in 1993 under the process outlined earlier, "would have provided for a shared stewardship body covering a larger area and having a somewhat different mandate and membership criteria."[25] Internal differences among the constituency of the aboriginal party, however, delayed ratification prior to the provincial election. Premier Mike Harris' Conservatives have since stated their intention to retreat from the agreement-in-principle.

Lack of provincial government political support for the WSA concept is compounded by, and probably reflected in, the financial difficulties faced by the Authority. Because it is kept on a short leash of annual budget allocations from the provincial government, the ability of the WSA to act autonomously is seriously compromised, and it has lacked the staff required to execute its mandate effectively.

The Royal Commission pronounces the following verdict overall:

> Whatever its future, there have been several positive lessons from this experiment in shared jurisdiction. Not only has the Wendaban Stewardship Authority generated support and collaboration among a multitude of often conflicting interests, at both the regional and local levels, it has also proven that Aboriginal and non-Aboriginal people can work together on issues of land and resource management. That in itself is a major accomplishment.

---

[21]Royal Commission on Aboriginal Peoples, *op. cit.*, 755.
[22] Jim Morrison, Wendaban Stewardship Authority, personal communication.
[23] *Op. cit.*, 754.
[24] *Op. cit.*, 754-5.
[25] *Op. cit.*, 755-56.

We are left with a puzzle. It is often claimed that the function of the state is to mediate diverse interests in society. But if diverse local and regional interests are able to resolve their differences, as the WSA experience indicates, then what are the provincial government's motives in frustrating constructive processes of this nature? And if the interests of people-on-the-ground are not being served, then who, exactly, is the state catering to?

Policies to delegate greater powers to the local level are easy commitments for government to make. However, questions remain unanswered regarding how much finance should be committed, how much leeway should be allowed for fiscal policy, and what legislative and regulatory powers should be ceded to local-level control. It is one thing to devolve responsibility to the local level. Without major additional resources or powers to support their activities, it is little more than tokenism.

Elk Lake Community Forest Project

Elk Lake is a lumber town 90 km north-west of Temagami with a population of about 550 people. As in many northern communities, the principal economic base of Elk Lake is a small sawmill. More than one hundred men are employed in the local Elk Lake Planing Mill, while the majority of the other residents are tied in some way to lumbering. This reliance on logging intensified in the 1970s and 1980s as a result of the 1973 Land Caution which prohibited all commercial development, with the exception of timber activity. Pressures from environmental lobby groups increased during the same period, putting further strain on the community as their way of life came under the public spotlight. Non-Native residents of these northern communities are inclined to feel that there has been little appreciation for *their* distinct and threatened lifestyles, with aboriginal and environmentalist concerns more readily gaining media attention.

The Elk Lake Community Forest (ELCF), an initiative under the Sustainable Forestry Program, is an attempt to reconfigure forest resource management through the promotion of local input and community development. The ELCF Partnership involves a land-base of about 470,000 ha, and includes representatives of forestry, industry, mining, tourism, environmentalist and labour interests. The Partnership aims to provide a forum for the various interest groups to meet and work out reasonable compromises. Feedback from the general public is also solicited. As a result, community members feel that they have obtained some power over resource management decisions that affect their lives and livelihoods, in contrast to the pre-existing system in which most major decisions were made elsewhere, and reflected broader provincial rather than local interests. In the words of Paul Tufford, manager of the ELCF:

> To my way of thinking this is the way it should go – to have local people getting involved in decision-making. A lot of decisions have been driven by people far outside the area, in Toronto or wherever, that affect the people who live and work and play here. When someone else was making the decisions, local people felt that their input didn't matter and there wasn't much incentive to get involved. Well, if you're making the decisions, then you're going to be involved and responsible. Most of the people want to live here and have no interest in seeing the place mowed down, or hunted and fished out.[26]

The ELCF seeks to accomplish more than the juggling of competing interests. It wants to demonstrate that sustainable forestry is a viable alternative that can accommodate both economic and environmental interests. Sustainable forestry is an approach that includes holistic concerns for wildlife habitats, tourism promotion, and public education. The ELCF project is also motivated by community development

[26] Paul Tufford, manager of the ELCF, as quoted in B. Griffin, What's the buzz? The Elk Lake Community Forest Project, *Highlander*, May/June, 1995, 13.

objectives. By offering training and employment for local people in silviculture work, habitat surveys, fish and animal data collection, and snowmobile trail work, the project creates much-needed employment opportunities in a region that has one of the highest rates of unemployment and where more families live below the poverty line than anywhere else in Ontario.[27]

Discussion

In recent years provincial and territorial governments across Canada have experimented with a range of models for increasing community involvement in land and resource management decisions. This trend toward the decentralization of planning and control is partly a response to pressure from residents of rural and remote communities who feel that local concerns about employment and access to resources have not been adequately represented. Other residents have argued that policies must give greater weight to non-extractive uses of natural resources. Aboriginal residents, who are found in both camps, feel that fundamental issues of ownership and jurisdiction have not been properly addressed. To all of these, co-management offers a process for working through conflicts at a grassroots level that have been intractable for central policy-makers. At the same time, central governments would seem to have a heightened incentive to devolve power in a period of fiscal restraint. Experience shows, however, that governments are more eager to cut program expenditures than to surrender legislative powers.

As we have seen from the three community-based approaches being taken in the Temagami region, the forms taken by decentralization and local empowerment vary greatly. Ontario currently applies the term "co-management" to a wide variety of stakeholder boards or committees, and in weaker forms the concept may function more as placebo than as remedy. In the Temagami region, co-management institutions range from advisory bodies such as the CPC to decision-making bodies such as the WSA. Even the latter case, however, falls well short of a real sharing of jurisdiction. Government control over operating budgets and other resources entails a power imbalance that may ultimately cripple the participation of local groups as equal partners with government, even when, as with First Nations groups, lip service is paid to the notion of shared powers.

Teme-Augama Anishinabai stewardship director, Mary Laronde cites the deep-seated reluctance of central governments to yield real control:

> When the ministry is becoming an advocate of co-management, both in process and in content (process more than content at this time, I think), the question is: Can it? Can that big machine, that big bureaucracy that has total control of all Anishinabai lands and the lands that we share with our neighbours and with the people that have come to live here, can it divest itself of that control in our interest? It protects its own interest in those lands. Can MNR become an agent for the public, for the people of Ontario? Or is it always going to be in control? That is a major issue. When you have that kind of entrenchment in both thought and power, in processes in thinking and in how things are done, the initial response is that the status quo must be maintained.[28]

To what extent are the interests of the broader Ontario and Canadian publics compatible with those of northern residents who depend most heavily and immediately on local resources? The issue of forest resource management is often misconstrued as a stereotypical opposition of (rural) logger versus (urban) environmentalist, of north versus south. Yet, in Temagami, a coalition of five environmental groups made

[27] B. Griffin, *op. cit.*

[28] M. Laronde, Co-Management of lands and resources in n'Daki Menan, in A-M. Mawhiney, ed. *Rebirth: Political, Economic, and Social Development in First Nations* (Toronto/Oxford: Dundurn Press, 1993), 101-2.

local control the central feature of their proposed land use plan. The Temagami case is clearly more complex than media reports would often have us believe. In the words of journalist, Linda Pannozzo:

> ...the issues do not break down over neatly defined lines like northerner versus southerner. A number of northerners have been involved and will continue to be involved in environmental issues regarding Temagami. In addition is the fact that crown land is owned by the people of Ontario. Like it or not, the public has a right to express their concerns over the management of these lands. With that right comes a responsibility to be diligent in seeking out information and developing a sense of historical context.[29]

What is the source of the provincial government's impulse to limit the sharing of power with local people, whether aboriginal or non-aboriginal? Is the government's role as defenders and mediators of the general "public interest" a smokescreen that enables it to cater to its own institutional needs, or those of powerful corporate interests?

It is not clear that anyone could benefit beyond the very near term from cutting into the tiny percentage of original white pine forest that remains. In theory, if such cutting were done, it should not occur at a rate faster than it is replaced by the emergence of new *old growth* areas of equivalent ecological complexity, which can only occur through a decades-long maturation process. There may be profits, for a short time, and additional jobs, for a few brief years only. The fact that remnant old growth areas are at risk of further reduction, despite their very high ecological, aesthetic and cultural value, is an indication of our reluctance to live within limits.

Questions

1. Based on the examples outlined in this case, what do you believe are the most important characteristics of community-based management approaches?

2. With the emergence of community-based management, is there a danger that the goals of conservation could potentially be compromised?

3. As local community groups, native and non-native, become more empowered, what role if any will advocate groups, particularly environmental groups, have?

4. The issue of forest resource management is often reduced by the media to a stereotypical opposition of (rural) logger versus (urban) environmentalist, of north versus south. What are some of the possible impacts of this, and how could they be avoided?

Further Reading

Berkes F. 1991. Co-management, *Alternatives* 18, 12-18.

Bray, R.M., and A. Thomson, 1990. *Temagami.* Toronto: Dundum.

Hodgins, B. W. and J. Benidickson. 1990. *The Temagami Experience: Recreation, Resources and Aboriginal Rights in the Northern Ontario Wilderness.* Toronto: University of Toronto Press.

Shute, J.J. 1993. *Co-management under the Wendaban Stewardship Authority: An Inquiry into Cross-cultural Environmental Values.* M.A. thesis, Carleton University, Ottawa.

---

[29] L. Pannozzo, Beyond the Great Divide: addressing the tension between north and south, *Highlander* Magazine, January/February, 1996, 32.

Web Sites

- Wendaban Stewardship Authority: Comments in the report of the Royal Commission on Aboriginal Peoples: http://www.libraxus.com/cgi-bin/folioisa/
- Canada and Agenda 21 on Forests: http://iisd1.iisd.ca/WORLDSD/CANADA/PROJET/c11.htm
- Elk Lake Community Forest: http://server1.nt.net/~elklake/elcf.html
- Forest Management Branch, Govt. of Ontario: http://www.gov.on.ca/MBS/english/programs/MNR1227.html
- Temagami Links: http://www.uoguelph.ca/~rolajos/temagami.html
- Lady Evelyn (Temagami): http://www.crca.ca/cca/sc17.html
- Temagami Land Use Zones Map: http://nrserv.mnr.gov.on.ca/MNR/temagami/map.html
- Owain Lake Forest, Indigenous Environmental Network: http://www.alphacdc.com/ien/a-owain2.html
- Greenpeace position: http://www.greenpeace.org/~comms/cbio/prfsep25.html
- Earthroots Campaign: http://www.earthroots.org
- Temagami News: http://server1.nt.net/~temagami/news

Audio-Visual Material

- *Battle for the Trees*, Sarus Productions Ltd., Otmoor Productions and the National Film Board, 1993 (dir.: John Edginton, 57 min.).
- *Temagami: The Last Stand*, Canadian Broadcasting Corporation, 1989 (47 min.).
- *Hewers of Wood: Out on a Limb*, Canadian Broadcasting Corporation, 1992 (17 min.).

Case Five
Atlantic Sealing: Immoral Slaughter or Sustainable Harvest?

Focus Concept

The role of science, politics and ethics in decisions relating to the sustainable utilization and management of a natural resource.

Introduction

Animal-rights activists are angered that the Canadian Government has increased the annual harp seal quota in eastern Canada. The Federal Department of Fisheries and Oceans (DFO) justifies the expansion of the hunt by blaming harp seals for eating Atlantic cod and destroying commercial fish stocks. Critics charge that politics is overruling the scientific facts: scientists say there's no evidence of a burgeoning seal population or that harp seals threaten cod stocks.[1]

History of the Seal Protest

Starting in the 1960s, considerable national and international opposition developed to the seal hunt off Newfoundland. Opponents alleged that the hunt was unethical and that the seal population had been seriously depleted. The protest was directed at the Canadian Government, who regulated the seal hunt, and the Norwegian Government, because Norwegian sealers accounted for a substantial share of the total harvest. Opposition to the seal hunt focused on the practice of "clubbing to death" harp seal pups and resulted in the European Economic Community banning the import of seal products in 1983.[2] The decline in demand effectively eliminated the seal hunt as a commercial hunt and led to the withdrawal of Norwegian sealers from Newfoundland.[3] Since 1983, harvests have been only a fraction of the total allowable catch. For example, in 1994 the total allowable catch for harp seals was 186,000, while the actual harvest totalled 60,000.

In May 1994, a parliamentary committee recommended expanding the seal harvest and giving government support to finding new markets for seal products. An internal government report, released in June 1995, indicated the extent to which the harp seal population is increasing, and estimated the level of consumption by harp seals of various fish species, particularly cod fish. In 1996, with the support of new Canadian Government subsidies, the biggest seal hunt since the early 1950s was undertaken, with more than 260,000 harp seals killed. The total allowable catch for 1997 has been set at 275,000.[4] As a result,

[1] It is interesting to note that advocates of sealing, such as DFO, usually refer to "the seal fishery", while opponents tend to use the term "seal hunt".

[2] Although the ban was limited to whitecoat seal products, the overall demand for seal-skins fell by approximately 50 percent, while the average per skin price fell from $25.80 in 1982 to $12.49 in 1983 and $10.00 in 1984 (A presentation by the Government of Newfoundland and Labrador to the Royal Commission on seals and the sealing industry in Canada, Department of Fisheries, St John's, May 1985).

[3] After the European ban, only 64,500 seals were taken in 1983, down from 153,000 in 1982 and from more than 190,000 in 1981 (A. Herscovici, *Second Nature: The Animal Rights Controversy*, Montreal: CBC Enterprises Radio-Canada, 1984, 71).

[4] Fisheries and Oceans Canada, Mifflin announces 1997 Atlantic seal management measures, News Release, NR-HQ-96-101E, 23 December 1996.

animal-rights groups – led by the International Fund for Animal Welfare (IFAW) - have renewed their campaign to force Canada to end the commercial hunt.

Biological Attributes

Six species of seal - the harp, hooded, grey, ringed, bearded and harbour - are found off the Atlantic coast of Canada. Harp seals (*Phoca groenlandica*) outnumber the population of all other seal species put together, and account for most of the harvest.

The harp seal is a marine mammal belonging to the suborder Pinnipedia (fin-footed animals) and the Family Phocidae (true seals lacking external ears). The name "harp" refers to the irregular horseshoe-shaped band of black across the back of the adult male. The rest of the pelt is steel blue in colour when wet and pale grey when dry. The head and tail are black, while the anterior flippers and belly are whitish. Adult females are similarly patterned, except the head, the "harp", and the tail are usually lighter in colour. Males are slightly larger than females; the average length of adult males is 169 cm, and of adult females 162 cm. Harp seals are long-lived and often reach ages of 30 years or more. Female harp seals become sexually mature between four to six years of age, males reach maturity at seven or eight years of age.[5]

Harp seals migrate every year between the Arctic and sub-Arctic regions of the North Atlantic. Three distinct stocks of harp seal have been identified in the North Atlantic, that in the Northwest Atlantic being the largest (Figure 5.1). These seals migrate to areas off eastern Newfoundland and in the Gulf of the St. Lawrence where they gather on the ice floes in late February or March to give birth (whelp).[6] Pups weigh about 11 kg at birth but quickly gain weight to reach 35 kg after about 10 days of nursing.[7] Within 10-14 days the pups are weaned and abandoned by their mothers. After leaving their pups, the females mate with the males and feed heavily on fish for several weeks before commencing their long journey northwards with the retreat of the pack ice. The pups begin to moult after weaning and are fully moulted at four weeks when they enter the water and begin to feed. Soon after they reach this stage, they begin to gradually move north to the main summer feeding grounds in the Canadian Arctic and off the coast of Greenland. Both adults and juveniles are most vulnerable to hunters during the period prior to their departure in March and April; the juveniles are easily clubbed to death before they become aquatic and the large concentrations of adults on the ice at this time make them easy targets.

Despite being subjected to the largest commercial hunt of any marine mammal in the world, the harp seal has managed to maintain relatively large populations, and although its numbers were reduced by hunting, it has never become endangered. The Royal Commission on Seals and the Sealing Industry, appointed by the Canadian government in August 1984 and comprised of marine-resource experts from Canada, Australia, Britain, and the United States, estimated the population of harp seals at the end of 1985 to be approximately 2-2.5 million.[8] More recent estimates, from the Department of Fisheries and Oceans, suggest that the total harp population in the Northwest Atlantic is 4.8 million, twice what it was in 1981, and the population continues to increase at an average rate of five percent a year, or about 250,000 seals.[9]

---

[5] W.D. Bowen, The harp seal. Underwater World Factsheet (Ottawa: Department of Fisheries and Oceans, 1991).
[6] P.A. Comeau, Sealing: a Canadian perspective. Underwater World Factsheet (Ottawa: Department of Fisheries and Oceans, 1989).
[7] Harp seals have one of the shortest known nursing periods of any mammal (*op. cit.*).
[8] A.H. Malouf, *Seals and the Sealing Industry in Canada*, Report of the Royal Commission (Ottawa: Supply and Services Canada, 1986).
[9] Department of Fisheries and Oceans, Report on the status of harp seals in the northwest Atlantic. Report 95/7 (Ottawa: DFO Atlantic Fisheries Stock Statistics, 1995).

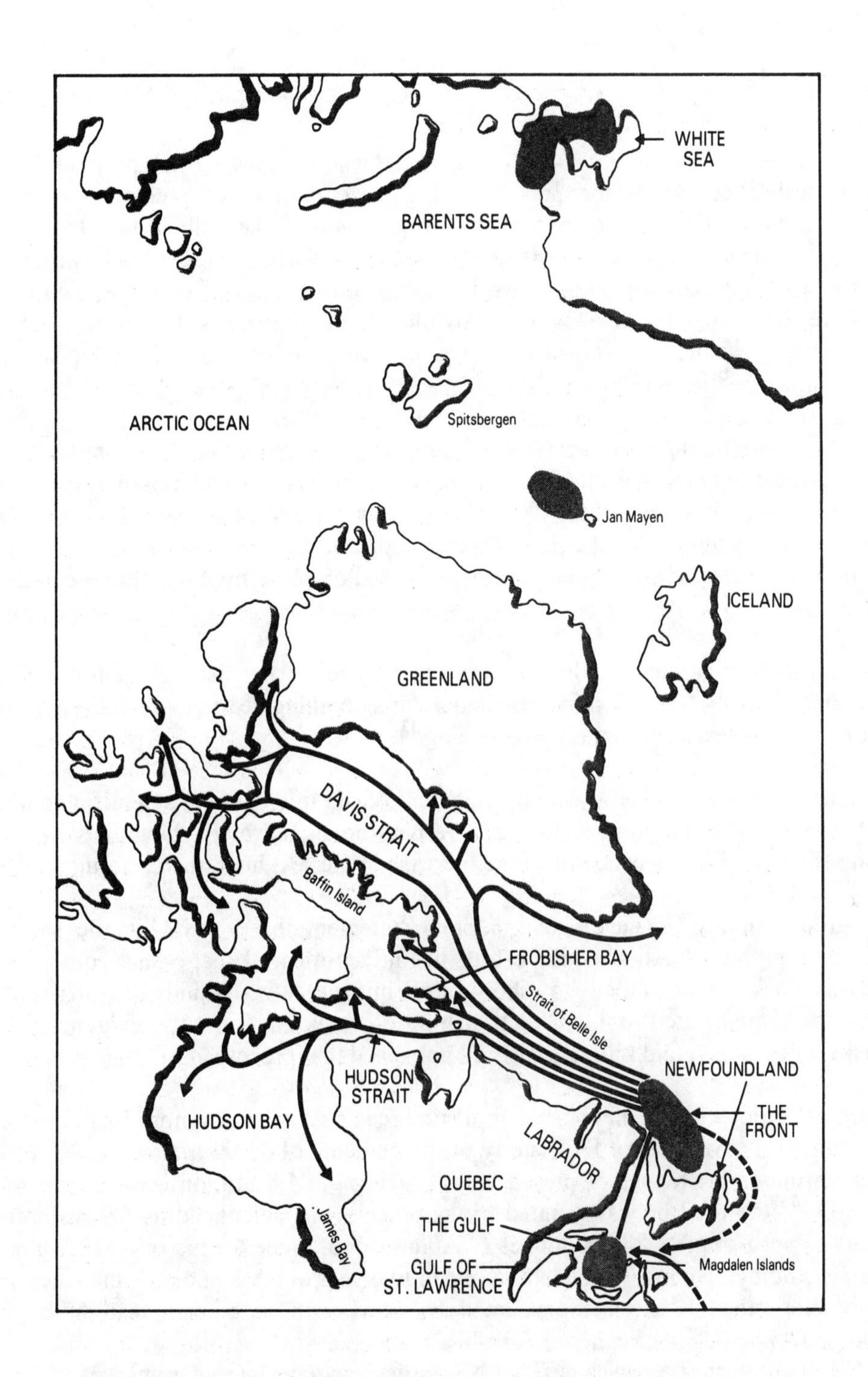

Source: Department of Fisheries and Oceans, 1989.

Figure 5.1. Breeding and Moulting Areas, and Principal Migration Routes of the Harp Seal Population.

Case Five

The Predator Pit

Biologists hypothesize that "depensation - a vicious circle where surviving fish produce fewer and fewer offspring - can occur because of the presence of a large number of predators which prevent the endangered species' population from recovering to former levels."[10] This situation is known in biology as the "predator pit". In the case of Newfoundland, the predator pit has been used as an argument "by those claiming that an exploding seal population must be culled so that decimated ground fish populations can return to former levels."[11] The image constructed by the Canadian Sealer's Association (CSA) and others of "voracious" and "ravenous" seals has been readily embraced by federal and provincial fisheries ministers in discourse on "the cod crisis".

At the same time, some research conducted by biologists provides evidence exonerating seals for the depletion of cod stocks. One such study is based on a review of the population dynamics of 34 commercially fished species spread over 128 of the world's overfished stocks.[12] No evidence was found in 125 of these stocks, including Northern cod, Baltic cod, and all other examples of fish similar to cod, that a depensation effect related to seals had occurred. Based on these findings, the researchers concluded that:

> Decimated fish populations like the northern cod will recover if fishing is cut down ... and what happened to the [East Coast] fish stocks had nothing to do with the environment, nothing to do with seals. It is simply overfishing.[13]

These conclusions have been backed by a large segment of the international scientific community.[14] The International Society for Marine Mammalogists, in a petition signed by 97 biologists from 15 countries and released in January 1996, has condemned the increase in the seal hunt quota, stating:

> We disagree with the Canadian government's statement that North Atlantic seals are a "conservation problem". All scientific efforts to find an effect of seal predation on Canadian groundfish stocks have failed to show any impact. Overfishing remains the only scientifically demonstrated problem ... If fishing closures continue, the evidence indicates that stocks will recover, and killing seals will not speed that process.[15]

Part of this group's critique is that an overly simplistic "fewer seals equals more fish" model is used by some DFO bureaucrats.[16] An internal DFO study of the contents of 5,000 seal stomachs, estimated that each harp seal consumes 1.4 tonnes of prey a year.[17] Using the 4.8 million estimate for the harp seal population in 1994,[18] the total prey consumed by harp seals was calculated as 6.9 million tonnes per year; this includes approximately 88,000 tonnes of Atlantic cod. Since the majority of fish consumed by seals are 10 to 20 centimetres long, equivalent to one to two year old Atlantic cod, the study inferred that there are many more fish in a tonne consumed by seals than in a tonne caught by commercial

[10] R.A. Myers, N.J. Barrowman, J.A. Hutchings, and A.A. Rosenberg, Population dynamics of exploited fish stocks at low population levels, *Science* 269, 1995, 1106-1108.
[11] S. Strauss, Decimated stocks will recover if fishing stopped, study finds, *Globe and Mail*, August 25, 1995.
[12] R.A. Myers *et al.*, *op. cit.*
[13] *Op. cit.*, 1108.
[14] R. Smith, A precious tradition? *New Maritimes*, May/June 1996.
[15] L. Hurst, Scapegoating seals, *The Gazette*, January 28, 1996, B4.
[16] R. Smith, *op. cit.*, 8.
[17] G.B. Stenson, M.O. Hammill, M.C.S. Kingsley, B. Sjare, W.G. Warren, and R.A.. Myers, DFO Atlantic Fisheries Research Document 95/20, 1995.
[18] P.A. Shelton, G.B. Stenson, B. Sjare, and W.G. Warren, Model estimates of harp seal numbers for the Northwest Atlantic, DFO Atlantic Fisheries Research Document 95/21, 1995.

fishermen.[19] Based on these results, the Canadian Government concludes that harp seals are one of the factors impeding groundfish stock rebuilding.[20] Because the harp seal harvest dropped from 300,000 animals a year in the mid-1960s to fewer than 65,000 in recent years, DFO claims that up to 287,000 seals could be harvested without reducing the current harp seal population. The fact that seals have few natural predators, except sharks, killer whales and polar bears, is also used to support this claim.

David Lavigne, a University of Guelph marine biologist, has studied the harp seal for 25 years. Lavigne's research indicates a more complex pattern of community interaction involving 80 species of plants and animals off the Newfoundland coast. A food chain chart shows that while cod make up three percent of the seals' diet, seals also eat the cods' other predators. Peter Meisenheimer, a research ecologist at the International Marine Mammal Association in Ontario, has made some calculations based on DFO's own figures to demonstrate that killing seals could actually be detrimental to cod stocks. His argument is based on the fact that some species of squid eat young cod, and seals eat squid. By reducing the number of seals, more squid survive to eat more cod.[21] In the absence of more detailed scientific data, Lavigne believes that it is impossible to know for certain whether the removal of seals would hinder or help the recovery of the cod.

DFO scientists observe that harp seals are reaching sexuality maturity later and having fewer young.[22] They suggest that these are signs that the harp seal population may be reaching its carrying capacity and stabilizing. Some analysts are critical of interpretations that infer the possibility of attaining a "balance of nature" once "unnatural" events, such as hunting by humans, are eliminated.[23] This type of reasoning assumes that in the absence of hunting, wildlife eventually reach a stable peak population and remain there. This is assumed to be the "carrying capacity" of the environment, "a magic factor which can be reached if populations are just left alone long enough to achieve it."[24] The reality according to Janice Scott Henke can be very different from this. Because environmental factors are constantly changing, wildlife populations must constantly change. Thus the assumption of a "natural" static carrying capacity is fundamentally flawed. According to Scott Henke, the seal population will regulate itself through natural controls once hunting stops. But these natural controls are not always "less cruel and stressful for the individual animals than death by hunting:"[25]

> The bottom line is that through chronic hunger, fertility of the herd is lessened and growth is numbers stops, then numbers decline. An environmental adjustment has been made, but at a cost of discomfort and suffering for millions of animals.[26]

These DFO scientists conducted other research on the size and appetite of the harp seal population, which was submitted as internal reports. These reports are "grey" literature as opposed to peer reviewed scientific publications, and all carry the caveat that these reports are "not intended as definitive statements on the subjects addressed but rather as progress reports on on-going investigations" and that "these estimates should be considered preliminary and used with caution." Despite this, an estimated population of 4.8 million harp seals in 1994 has been widely quoted without explanation or context, and provides the

---

[19] In 1992, one tonne of commercial catch of Atlantic cod accounted for 900 fish. One tonne of one-year Atlantic cod consumed by seals is equal approximately to 38,000 fish, while one tonne of two-year old Atlantic cod is about 9,500 fish.

[20] Department of Fisheries and Oceans, Seals abound in Canadian waters, Understanding the Seal Fishery, No. 3, Cat. No. Fs 41-42/1-1995E.

[21] *Op cit.*

[22] P.A. Shelton et al., o*p. cit.* and G.B. Stenson et al., o*p. cit.*

[23] J. S. Henke, *Seal Wars: an American Viewpoint* (St. John's, Nfld.: Breakwater, 1985)

[24] *Op. cit.*, 79.

[25] *Op. cit.*, 78.

[26] *Op. cit.* 78.

basis for the government's decision to increase the seal quota.[27] According the Lavigne, there are serious scientific problems with the study related to "questionable statistical analyses and the incorporation of uninformed and inappropriate biological assumptions."[28] The 4.8 million estimate comes from a computer model that incorporates earlier, incompatible population estimates, obtained by different research designs and methods (mark recapture as opposed to aerial survey). Furthermore, the model is based on the biologically unrealistic assumption that natural mortality is constant for animals of all ages (i.e. the death rate of very young seals is equal to that of older animals).

Peter Meisenheimer believes that we would know more about the population dynamics and complex interactions of this marine community if the federal government had not cut back globally respected laboratories studying marine ecology and laid-off many of the Department of Fisheries and Oceans' research staff. He is concerned that political decisions continue to ignore modern environmental thinking, scientific data and common sense:

> The cod crisis isn't caused by voracious seals – it has its roots in international relations that do not include arrangements for protecting the environment and economic policies that don't recognize the limits of natural systems. It was accelerated, of course, by opportunistic politicians whose arrogance is so absolute that they chose to play God.[29]

As things stand there is much uncertainty about the numbers of harp seals and nobody knows whether culling seals would have an impact on the cod, let alone the direction of that impact.

Political Considerations

If the scientific community at large says that there is insufficient evidence to indicate that seals are obstructing the cod's return, why has the seal hunt quota been enlarged? To animal-rights critics the answer is politics; political considerations are overruling scientific facts. They charge that the seal hunt has been expanded to placate economically depressed Newfoundlanders. Those who benefit from the use of the resource on the other hand, counter that there is nothing wrong with that; and that the scientific evidence clearly demonstrates the feasibility of a sustainable harvest, whether or not seal culls are needed to assist cod recovery.

Fred Mifflin, Minister of Fisheries and Oceans, stated in a recent news release:

> The seal fishery provides much needed employment and income for residents of coastal communities devastated by groundfish moratoria.[30]

There is little question that thousands of people have a stake in the East Coast seal fishery. Besides the hunters there are the equipment suppliers, truckers, wholesalers and processors. Sealing is a particularly

[27] One of the greatest challenges of wildlife management is related to the difficulty of estimating population size. Adult seals are difficult to count because many are underwater. Instead an estimate of the number of pups born in the population is obtained by aerial survey during the whelping period when they remain on the ice. Logistical difficulties related to the size of the nursery area and cost of the surveys mean that only 2 to 7 percent of newborns are actually counted. A statistical averaging technique is then used to estimate the entire pup population. The total seal population is then calculated by multiplying the pup population by a number ranging from 4 to 6.5, depending on the number of new breeding seals in a given herd – the result is a harp seal population of 4.8 million, ± 2 million.

[28] D.M. Lavigne, Comments on Canada's "Report on the status of harp seals in the Northwest Atlantic," Unpublished paper (Guelph: International Marine Mammal Association Inc., n.d.).

[29] P. Meisenheimer, If every seal in the Atlantic became a vegetarian ..., *Globe and Mail*, March 6, 1992.

[30] Fisheries and Oceans Canada, Mifflin announces 1997 Atlantic seal management measures, News Release, NR-HQ-96-101E, December 23, 1996.

important source of income for the fishermen who live in small, isolated communities where employment opportunities are scarce and seasonal. In many cases, catching the seals is the only way to raise money to get their boats and equipment seaworthy for the summer fishing season. Indeed the Royal Commission into the Canadian sealing industry acknowledged:

> The loss of income from sealing weakens the whole annual cycle of activities and thus threatens the survival of some of these communities. Alternatives to sealing have been considered by the Royal Commission but the prospects are not good.[31]

Market Factors

Critics of the expanded seal hunt suggest that a further misconception of the hunt is that a viable commercial market exists to justify the increased quota.

According to the DFO, economic benefits from the 1996 seal harvest were around $11 million. They claim that several meat products are currently being tested for their market potential, while others are already being marketed for animal feed. Interest is also said to be growing in seal oil as a source of pharmaceuticals; the blubber is rich in omega three fatty acids which medical research has indicated can play a significant role in preventing many common cardio-vascular diseases. Marketing initiatives for fur and leather products are also taking place in Canada, China, Japan, Korea, Norway and Denmark. However, the greatest potential market for any seal product exists in parts of Asia, where the seal penis is prized as an aphrodisiac. To discourage the harvesting of seals exclusively for individual components, such as penises, the Canadian Government has introduced a policy of full utilization of harvested seals through the provision of a meat subsidy.

Critics question why, in the presence of a so-called viable commercial market, financial assistance to the sealing industry in Newfoundland has been set at $750,000 for 1997; this figure includes a subsidy of 35 cents for every pound of landed seal meat and a subsidy of $130,000 to support the Canadian Sealer's Association. The "expanding market" was one factor cited by former Fisheries Minister Brian Tobin when he increased the 1996 total allowable catch of seals to 250,000. Yet in the previous year, the quota had been set at 186,000 seals, although just over a third of that number were, in fact, harvested.

The results of a marketing study conducted in 1995 for the Northwest Territories seal industry raise further concern about the economics of the hunt. The study found only one product, seal penises, to have "excellent" economic prospects.[32] The market for seal meat, both for human and animal consumption, was rated poor, as was fur apparel and seal oil for industrial use. Seal oil for health products and seal leather for accessories were rated as having potential, with the proviso that both faced high development costs and the latter "marketing problems" because of the ban on many seal products by European Union countries and the United States.

The IFAW has taken a different approach to the marketing issue by encouraging the development of a seal watching tourism industry; they hope to prove that the seals are worth more alive than dead.

---

[31] A.H. Malouf, *op. cit.*

[32] An Asian syndicate is reported to have received a permit from the Newfoundland Government to export seals. The syndicate wanted to buy 60,000 penises and was offering to pay $50 each (Anon, Atlantic storm brewing over seal penises, *Globe and Mail*, November 6, 1993, A3).

Regulating the Hunt

The Report of the Royal Commission, presented in 1986, found that the seal hunt was a legitimate activity, which should be continued subject to the principle of sound management of the stocks. In response to the 45 recommendations of the Commission, the Canadian Government announced a new seal policy in December 1987. Under the new policy the commercial harvesting of young harp (whitecoats) and hooded (bluecoats) seal pups was banned, and the large-vessel offshore seal hunt was no longer permitted. As a result, the current seal hunt is an inshore activity, carried out by rural and coastal inhabitants, which focuses on juvenile and older seals.

About 8,900 commercial licences were issued in eastern Canada in 1994. To encourage professionalism, these licences were limited to registered professional fishermen. Personal Use Licences have been introduced to help those East Coast fishermen in areas hard hit by fisheries closures. Seals are an important source of food for residents of these areas. Hunters with Personal Use Licences are allowed to take up to six seals for their own use but are prohibited from taking younger animals. In 1995, less than 900 licences were issued and the harvest totalled 1000 seals. East Coast Inuit and Indians who harvest seals on a non-commercial basis, using skins for clothing and meat for food, are exempt from the licensing requirement. Recreation or "sports" hunting for seals is not permitted.

Ethical Considerations

In addition to the ecological, economic and political issues raised above, film footage and photographs, obtained by anti-sealing groups, have made ethical considerations a central issue in the sealing debate. Images of baby seals being clubbed to death became a familiar sight throughout the earlier campaigns launched by Greenpeace and IFAW. Film footage has shown that during the rush to kill whitecoat seals, some animals were only stunned by an inadequate clubbing, and then skinned while still alive. Apart from allegations that some of these incidents were deliberately staged by the anti-sealing groups, the film footage has been criticized for falsely portraying seals that appear to be "alive" even after death. Because seals spend a great deal of time under water, their blood is capable of storing large amounts of oxygen. A seal's muscles can also function for a while without a supply of oxygen. As a result a seal's body can display powerful spasms for a long time after its death.

The Royal Commission, as well as qualified veterinarians, animal pathologists and biologists who have observed the hunt first-hand have attested to the humaneness of the clubbing method when it is carried out properly. Indeed the Commission went as far as saying that the harvesting methods used are even more humane than some practices in commercial slaughterhouses. A 1993 report from the Parliamentary Assembly of the Council of Europe also found that "clubbing", if correctly carried out, is as good as the usual methods of slaughter, causing the animal to die in the course of a few seconds.

With regard to the use of firearms, the Report of the Royal Commission noted that as many as three out of every four seals shot in the water may never be recovered. They suffered and died, but were not counted in the reports made by the Department of Fisheries and Oceans. To address this problem, federal regulations, which specify the calibre of firearms and types of ammunition, were introduced to ensure that the seals are killed as rapidly and as humanely as possible. The regulations are also designed to ensure that only experienced hunters, fully able to comply with all the regulations, can participate in the hunt. With the hunt now focusing largely on older animals, firearms are used to a greater extent and the effectiveness of these regulations and controls needs to be closely monitored.

IFAW Involvement in the Seal Protest

The issue of the Atlantic seal hunt is a familiar one to the IFAW - it was the slaughter of baby harp seals that first motivated IFAW's founder Brian Davies to speak out against cruelty to animals. He began his campaign in 1965 and founded IFAW four years later to end the annual commercial hunt of more than 200,000 whitecoat seals in Canada. The Canadian Government gave in to international pressure in the late 1980s and banned the slaughter of the youngest baby seals. Since the ban, about 60,000 seals have been harvested on a yearly basis.

In late 1996 in the wake of calls for an expanded hunt, the IFAW announced that it would renew its campaign to pressure the Canadian Government to abolish the annual seal hunt. The campaign centered on a run of newspaper ads within Canada and in Britain. One approach by the IFAW has been to highlight the detrimental impact the "annual slaughter of seals" has on Canada's image abroad. In the words of IFAW spokesman, Nick Jenkins:

> Canada is a beautiful country, with so many good things going for it, but internationally the one massive black mark against it is sealing. It has been for the past quarter of a century. And it's time that it recognized the damage that it can do to its image and indeed to its tourism, and its other business interests is just not worth it.[33]

Another strategy has been to publicize illegal sealing activities. IFAW investigators have uncovered evidence that baby seals are illegally being killed for a new black-market trade for furs. Officials from the Department of Fisheries and Oceans seized more than 25,000 whitecoat and bluecoat pelts (estimated value $800,000) in November 1996, and 101 Newfoundland sealers were charged with illegally selling pup pelts to a local processing plant. The sealers received fines of up to $100,000 and licence suspensions.[34] The IFAW also wants charges laid against Newfoundland sealers engaging in apparently cruel acts filmed during last year's hunt. The scenes include seals being pulled from the water with large hooks, struggling after being shot non-fatally, and lying in bloody piles after being skinned. The IFAW obtained the video footage by hiring investigators who posed as photographers for a United States hunting magazine.

The tactics used by IFAW, Greenpeace and other anti-sealing groups have been the target of much criticism over the past three decades. The seal industry claims that anti-sealing groups have used the photogenic qualities of new-born harp seals to shift the focus of the debate from an issue of rational utilization of a natural resource to an emotional appeal. Others claim that environmentalists who scream about cute seals have nothing to say about the rights of cod.

Closing Thought

Both government departments and ENGOs have for many years produced research related to several relevant factors that have been largely ignored by government bureaucrats and the media. The focus on the seal hunt rather than on broader ecosystemic connections has been dictated as much by journalists and politicians as by the priorities of industry, resource users, or ENGO researchers and spokespersons. In the popular media, it is easier to sell superficial images than thoughtful analysis. This hinders the resolution, in public policy, of this complex and controversial issue.

---

[33] H. Branswell, Group launches new anti-sealing campaign, *Edmonton Journal Extra*, 11 November 1996.
[34] M. MacAfee, Officials charge 101 sealers in pup-pelt trade, *The Hamilton Spectator*, 20 November 1996.

Questions

1. Attention must be paid to multi-species interactions if harp seals are to be managed as one of several important predators in an ever-changing marine community. Outline the kind of research data that is needed.

2. In Canada, the same federal government department that is responsible for looking after fish stocks, the Department of Fisheries and Oceans, is also charged with the management of marine mammals. What benefits and problems are likely to result from this administrative arrangement?

3. The quality of scientific evidence varies considerably. By what criteria can we assess the reliability of scientific data and scientific "proof" in evaluating a controversial issue?

4. Is there a moral difference between killing wild creatures such as seals compared with the slaughter of livestock specifically bred, raised and killed for human consumption?

Further Reading

Busch, B.C. 1985. *The War against Seals: A History of the North American Seal Fishery*. Kingston: McGill-Queen's University Press.

Candow, J.E. 1989. *Of Men and Seals: A History of the Newfoundland Seal Hunt.* Ottawa: Studies in Archaeology, Architecture and History.

Herscovici, A. 1984. *Second Nature: The Animal-Rights Controversy.* CBC Enterprises Radio-Canada.

Lavigne, D.M. 1992. Interactions between marine mammals and their prey: unravelling the tangled web. *Symposium on Marine Birds and Mammals in Arctic Food Webs.* St John's: Memorial University of Newfoundland, 3-7 April.

Malouf, A.H. 1986. *Seals and Sealing in Canada.* Report of the Royal Commission, Ottawa: Supply and Services Canada.

Scott Henke, J. 1985. *Seal Wars: An American Viewpoint.* St John's: Breakwater.

Wenzel, G. 1991. *Animal Rights, Human Rights: Ecology, Economy, and Ideology in the Canadian Arctic.* Toronto: University of Toronto Press.

Web Sites

- Seals and seal-hunting: http://odin.dep.no/ud/nornytt/unn-141.html
- Brigitte Bardot Foundation: http://www.foundationbrigittebardot.fr/uk/reprise.html
- Canada Site: the hidden atrocities: http://www.kemptown.co.uk/animact/canada.html
- Department of Fisheries and Oceans: http://www.ncr.dfo.ca/communic/seals/understa/utsfl_e.html
- Siksik, learnet: http://siksik.learnnet.nt.ca/Inuuqatigiit/SealsK-3.html
- Siksik learnet: http://siksik.learnnet.nt.ca/Inuuqatigiit/Seals7-9.html

Audio-Visual Material

- *Canada's Seal Hunt: It's Not Just Cruel, It's Criminal*, Parador Television Communications, International Fund for Animal Welfare, 1996.
- *Pelts: Politics of the Fur Trade*, National Film Board of Canada, 1989 (dir.: Nigel Markham, 57 min.).
- *Sea of Slaughter,* Canadian Broadcasting Corporation, 1989 (95 min.).
- *Taking Stock*, National Film Board of Canada, 1994 (dir.: Nigel Markham, 48 min.).

Case Six
Sustainable Agriculture and Biodiversity Conservation in the Prairie Provinces

Focus Concept

Sustainable agriculture[1] on the prairies brings benefits through higher crop yields and improved soil quality,[2] as well as enhanced agro-ecosystem diversity.

Introduction

Although agriculture is often charged with the degradation and loss of wildlife habitat, agricultural use accounts for only seven percent (67.8 Mha) of Canada's total land area. The total area of farmland has remained relatively stable over the past twenty-five years suggesting that those lands suited to farmland development have been maximized and further expansion into non-agricultural land beyond the agricultural land base is unlikely.[3] Further increases in cropland must therefore come from changes in land use within the agricultural land base. A trend toward more intensive agricultural land use is already apparent in the Prairie Provinces (Manitoba, Saskatchewan, and Alberta), where the proportion of prairie land in summerfallow[4] decreased by about thirty percent since the early 1980s, while the cropland proportion of total farmland increased by 20 percent.[5]

Despite the limited land area devoted to agricultural land use in Canada as a whole, there is no doubt that agriculture has had a major impact on biodiversity at both species and ecosystem levels.[6] Eighty-five percent of the decline in Canada's original wetland area is related to drainage for agriculture,[7] while other original habitats, such as native prairie grasses, have suffered widespread conversion to agricultural use.[8] Substantial losses in biodiversity are associated with these changes. Agriculture can also contribute to the conservation of biodiversity. Practices such as the planting of shelterbelts, and maintenance of rangelands and woodlots, in addition to improvements in soil and water management through conservation tillage and low disturbance seeding (no-till) can have significant benefits for wildlife.

---

[1] Sustainable agriculture may be defined as land stewardship which permits the continued production of foods and fibres indefinitely, for the general benefit of society, while preventing the degradation of soil, water, air, and biological resources.

[2] Although sustainable agricultural practices can also lead to substantial improvements in water quality, these are not discussed in this case study.

[3] Statistics Canada, *Trends and Highlights of Canadian Agriculture and its People,* Catalogue No. 96-303E (Ottawa: Agriculture Division, 1992).

[4] Summerfallow refers to land that is not cropped for at least one year.

[5] Statistics Canada, 1996 *Census of Agriculture: National and Provincial Highlights* (Ottawa, May 1997).

[6] While biodiversity is used as a general term to mean the variety of life on the Earth, technically the concept consists of three components: 1) genetic diversity – the total number of genetic characteristics; 2) species diversity; and 3) habitat or ecosystem diversity – the number of kinds of habitats or ecosystems in a given unit area. This case study is primarily concerned with habitat/ecosystem diversity.

[7] C.D.A. Rubec, Canada's federal policy on wetland conservation: a global model, in W.J. Mitsch, ed., *Global Wetlands: Old World and New* (Amsterdam: Elsevier, 1994), 909-917.

[8] D.A. Gauthier and J.D. Henry, Misunderstanding the Prairies, in M. Hummel, ed., *Endangered Spaces: the Future of Canada's Wilderness* (Toronto: Key Porter Books, 1989), 183-195.

Case Six

The conservation of agro-ecosystem biodiversity is part of Canada's commitment to the United Nations Convention on Biological Diversity, the Canadian Biodiversity Strategy, and various federal and provincial action plans, among other initiatives. Since eighty-two percent of Canada's farmland is located on the prairies, agricultural developments in that area will have a major impact on the overall direction of change for issues related to the sustainability of agriculture and the conservation of biodiversity. This case study examines a number of key developments in these areas.

Agricultural Land Use in the Prairies

Agriculture in the Prairie Provinces occurs primarily in the Prairie ecozone,[9] an area comprised mainly of semi-arid grasslands and sub-humid aspen parkland. Farming also takes place in the southern part of the Boreal Plain ecozone. This area of cool, sub-humid, aspen forest extends from the Peace River region of B.C. to central Manitoba.[10]

The natural fertility and good moisture holding capacity of *chernozemic* soils has made the Prairie ecozone highly productive for crops. The relatively flat topography is suitable for highly mechanized farming, and although the growing season is relatively short at 110 days compared to other agricultural areas, early maturing varieties permit the growth of hard spring wheats, as well as other cereals, pulse crops and oilseeds. Beef cattle ranching occurs on sandier and more hilly areas. The Boreal Plain ecozone is associated with *luvisolic* soils which are more acidic and less fertile than those of the Prairie ecozone. Crop production is limited to short season cereals, such as barley, cool-season oilseeds like canola, and forage crops.

Although some farming took place along rivers in the early 1800s, most land settlement in the Prairie and Boreal Plain ecozones took place between 1880 and 1920. Land was usually settled on the basis of its proximity to rail lines, as much as its suitability for agriculture, and settlers introduced farming practices that were better suited to the moist conditions of eastern Canada and Europe.

Prairie soils were rich in organics and fairly resistant to wind erosion when first worked. The development of adapted wheat varieties, combined with good soil fertility and relatively moist weather resulted in high crop yields in the early 1900s. Within twenty years, however, inappropriate farming methods and drought had lead to serious problems with soil erosion and low crop yields. Droughts, devastating soil erosion and economic depression followed, resulting in a widespread abandonment of farms in the 1930s. Despite improvements to dryland farming practices in the 1940s and 1950s, soil health continued to deteriorate over the next couple of decades, especially in areas of extensive summerfallowing.[11]

Since the 1950s, the number of prairie farms has decreased, due to economies of scale which favour larger farm sizes.[12] In Saskatchewan, for example, the number of farms dropped from 112,000 in 1951 to 60,000 in the early 1990s.[13] This trend has been paralleled by an increase in crops and improved crop

---

[9] Ecozones are areas that have distinct climate, vegetation, geology, and soils.

[10] Agriculture and Agri-Food Canada, *Profile of Production Trends and Environmental Issues in Canada's Agriculture and Agri-Food Sector* (Ottawa: Minster of Public Works and Government Services, 1997).

[11] D.F. Acton and L.J. Gregorich, *The Health of Our Soils: Towards Sustainable Agriculture in Canada*, Publication 1906/E (Centre for Land and Biological Resources Research, Agriculture and Agri-Food Canada, 1995).

[12] The rural population of the prairie provinces declined from about 1.7 m in 1941 to 1.2 m in 1991, while the urban population increased from about 0.6 m to 3.4 m. Of the rural population, the on-farm portion has declined from about 1.1m in 1946 to about 0.4 m in 1991 (C. Bradley and C. Wallis, *Prairie Ecosystem Management: An Alberta Perspective*, Prairie Conservation Forum, Occasional Paper Number 2, 1996).

[13] D.F. Acton and L.J. Gregorich, *op. cit.*, 15.

varieties, higher crop yields, increased use of fertilizers and pesticides, and larger, more mechanized farms. While total farmland area on the prairies has increased by only two percent (55.4 Mha in 1991) over the past twenty-five years, land under crops increased by 25 percent.[14] Improved pasture land and unimproved land for pasture increased by 44 percent and four percent, respectively over the same period,[15] while the area in summerfallow decreased by 27 percent. Table 6.1 shows that these trends have continued in more recent years.

Table 6.1. Changes in Prairie Agriculture Land Use between 1991 and 1996 (hectares).

	1991	1996	% change
Land:			
Land in crops	27,193,464	28,312,688	4.1%
Summerfallow	7,691,129	6,120,112	-20.4%
Total pasture land	16,785,122	16,671,442	-0.7%
All other land	3,090,304	3,583,214	16.0%
Major crops:			
Total hay and fodder crops	10,447,939	12,347,526	18%
Canola	9,516,952	10,475,160	10%
Barley	7,686,544	8,585,707	12%
Oats	4,005,972	5,123,467	28%
Land management:			
Area irrigated	538,756	628,778	16%
Area applied with commercial fertilizer	17,487,916	20,635,69	18%
Area treated with herbicides	18,759,870	20,087,332	7%
Area tilled retaining most of the residue on surface (conservation tillage)	6,417,169	7,909,688	23%
Area tilled incorporating most of the residue into soil (conventional tillage)	16,718,802	12,756,293	-24%
Area of no-till	1,792,439	4,032,062	125%

Source: 1996 Census of Agriculture National and Provincial Highlights (Ottawa: Statistics Canada, 1997).

Farming Practices

The primary agents of soil disturbance in native grasslands in the past were probably bison, ground squirrels and badgers. Today, cultivation is the dominant agent. Conventional tillage breaks down soil structure and increases the risk of wind and water erosion. Tillage machinery also causes tillage erosion and compaction of the soil. The biggest factor affecting soil erosion, however, is the loss of ground cover due to burial of crop residues by tillage (Blair McClinton, personal communication). Some traditional cultivation practices also result in erosion and the loss of soil organic matter. These include up- and down-slope cultivation, monoculture, and fallowing.

Practices that enhance soil health do so by building up and protecting soil organic matter and soil structure. These include:[16]

- conservation tillage, including no-till
- residue management

[14] Statistics Canada, *op. cit.,* 1992.

[15] Increased pasture area can be partly attributed to enrolment in the Permanent Cover Program in the early 1990s.

[16] D.F. Acton and L.J. Gregorich, *op. cit.*, 117.

- contour cultivation on hilly land
- application of organic amendments such as manure, compost and sewage sludge
- reducing fallow by extending crop rotations or cropping continuously
- including legumes and forages in crop rotations
- water management
- structural erosion controls, including growing forages in rotation, interseeding, planting shelterbelts, strip-cropping, and restructuring the landscape (terracing, grassed waterways, diversions).

Conventional tillage buries crop residues in the soil to prepare a "clean" seedbed for the next crop. Conservation tillage, on the other hand, refers to methods of tillage that maintain a cover of crop residue (i.e. "stubble" or plant material remaining after a crop is harvested) on the soil surface, where it protects against soil erosion. Conservation tillage can either reduce the amount of tilling (reduced or minimal tillage) or eliminate it altogether ("no-till" or "zero till"). No-till has been described as an annual cropping system that mimics, in certain respects, the original native grass ecology of the prairies.[17] The permanent root complexes that it fosters resemble native sod in regard to soil structure, humus depth, and water retention capacity. The new crop is planted directly into the residue of the previous year's crop, offering the maximum residue cover of soil. The residue slowly decays on the surface, recycling nutrients into the soil for use by future crops The old roots plus the new ones anchor the soil and provide permanent aeration and water channels. Although herbicides and fertilizers are still used, they are required at lower levels than with conventional summerfallow or continuous cropping regimes.

Summerfallow has, traditionally, been used to conserve soil moisture in the driest areas. Although it produces short-term benefits in terms of higher yields than continuous cropping, summerfallow increases the risk of soil loss, salinization, and structural degradation. Evidence is accumulating that no-till systems have greater ability to enhance the longer-term moisture retention of soil through the development of spongy root masses, deeper humus levels, and improved soil structure (Jim Halford, personal communication). Thus, conservation tillage (including no-till or zero till) through improved soil quality can result in positive economic benefits as well as improved prospects for long-term sustainable production. However, even no-till options are incapable of reproducing the degree of biodiversity present in native grassland. Table 6.1 shows the relative increase in conservation tillage and no-till practices compared to conventional tillage in recent years.

Since ground cover is a major factor in soil erosion, crop choice and sequencing affect environmental risks. Different crops provide varying amounts and types of residue cover on soil, while important annual row crops afford less protection to soil (via residue cover) than perennial forage crops or most small grain crops. Incorporating winter cover crops, together with low tillage approaches and other soil conservation practices into cropping systems, is an effective method of reducing soil degradation risks.

Erosion controls include the planting of shelterbelts to protect soil from wind erosion. Crop damage by strong winds can be reduced, and soil moisture improved. Other benefits include farm diversification, and farmyard protection and beautification. There is some evidence that shelterbelts also modify weather in adjacent fields, creating a micro-climate that favours crop production. Average daily temperatures are increased, soil temperatures are higher and wind damage to crops is reduced drastically. These factors combine to increase yields of almost all crops protected by a shelterbelt system. Tree plantings also provide environmental benefits through their potential to sequester atmospheric carbon,[18] and through the

---

[17] J. Halford, Our no-till experience (Indian Head, Sask.: Vale Farms Ltd., 1994), 9

[18] J. Kort and B. Turnock, *Biomass Production and Carbon Fixation by Prairie Shelterbelts: A Greenplan Project*, Agriculture and Agri-Food Canada, PFRA Shelterbelt Centre, Supplementary Report 96-5, 1996.

provision of wildlife habitat. Table 6.2 shows the most commonly used methods of conserving soil erosion in 1996.

Table 6.2. Soil Conservation Practices in the Prairie Provinces, 1996

	Number of Farms	Percentage of Farms
Total farmland	139,305	
Crop rotation	92,429	66%
Permanent grass cover	40,222	29%
Winter cover crops for spring plough-down	1,923	1%
Contour cultivation	8,567	6%
Strip-cropping	8,433	61%
Grassed waterways	13,281	10%
Windbreaks of shelterbelts	24,569	18%

Source: 1996 Census of Agriculture National and Provincial Highlights (Ottawa: Statistics Canada, 1997).

Environmental Significance of Agricultural Land Use

Although prairie soils are inherently of good quality, many agricultural soils in the region are subject to the stresses of a dry climate and are susceptible to degradation, particularly wind erosion and salinization. As shown in Table 6.3, the extent of these degradative processes depends on the type of agricultural land use.[19]

Future environmental pressures are considered likely to arise from a variety of sources: the continued dominance of monoculture cropping systems, the increased area seeded to annual crops, and the expansion of intensive livestock operations.[20] Monoculture, while inherently unstable, gives high annual yields, creates its own economies of scale and uses specialization to increase production levels. Maintenance of the productivity of these lands will, however, require increasingly higher inputs. The recent increase in prices for wheat and other small grains is likely to result in an increase in annual cropland. This increase could lead to reduced soil quality, loss of wildlife habitat, and increased greenhouse emissions if it cuts into marginal land or reduces the area in forage crops. Alternatively, environmental impacts and risks could be reduced if the trend is towards an increase in area under small grains. This could bring about a further reduction in summerfallow and reduce the area seeded to oilseed crops, such as canola, which leave less protective residue cover after harvesting.[21]

The magnitude of these impacts will depend on whether inherent soil and landscape limitations are respected; whether soil nutrients are replaced through fertilization; and whether soil conservation practices, most importantly conservation tillage, and other erosion control practices are used.

Agro-Ecosystem Biodiversity

There has been a growing recognition in recent years within the agriculture community of the importance of maintaining healthy agro-ecosystems and conserving and enhancing wildlife habitat. Efforts to enhance biodiversity range from initiatives to conserve genetic material, to participation in endangered species protection programs. Biodiversity benefits are also derived from programs and farm practices aimed

[19] D.F. Acton and L.J. Gregorich, *op. cit.*

[20] Agriculture and Agri-Food Canada, *op. cit.*

[21] Canola production has increased dramatically in recent decades, and accounted for more than one third of the area under crops in the Prairie Provinces in 1996.

primarily at soil and water management. The following sections outline some of the initiatives currently being taken on the Prairies to conserve wildlife habitat.

Table 6.3. Environmental Significance of Agriculture Land Uses.

Land Use:	**Definition:**	**Environmental Significance:**
Total farmland	Sum of all land owned or rented by farmers	Depends on specific land use
Cropland	Total area of field crops, fruits, vegetables, nursery products, and sod.	Most intensively used and productive land, relatively greater use of pesticides and fertilizers, higher risk of soil degradation, less suitable as habitat.
Summerfallow	Land that is not cropped for at least one year	Higher risk of soil erosion, organic matter loss (oxidation), sedimentation of waterways, lower suitability for wildlife.
Improved pasture	Pasture area improved by seeding, draining, irrigating, fertilising, and brush or weed control, not including area where hay, silage, or seeds are harvested.	Less intensively used than cropland but more intensively used than unimproved pasture, greater and virtually continuous soil cover, low risk of soil degradation, more suitable as habitat for some species, some application of fertilizer.
Unimproved pasture	Area of native pasture, native hay, rangeland, grazable bush, etc.	Least intensive agricultural land use, greater and continuous soil cover, virtually no risk of soil degradation, higher value as habitat for some species, no application of fertilizer or pesticides.[22]
Other land	Farmland area occupied by farm buildings, lanes, woodlots, bogs, marshes, brush, improved idle land, tree windbreaks, etc.	Depends on specific land use.

Source: Agriculture and Agri-Food Canada, *Profile of production trends and environmental issues in Canada's agriculture and agri-food sector.* Catalogue No. A22-166/2 (Ottawa: Minister of Public Works and Government Services), 1997: 13.

Prairie Wildlife Habitat

More than 60 percent of prairie farmland was classified as unimproved land (implying higher quality habitat) in the early 1900s.[23] This proportion of unimproved land progressively decreased over subsequent decades to reach a low of 28 percent in 1981. An increase in the area of cropland since then has mainly resulted from a corresponding decrease in area summerfallowed rather than from the conversion of unimproved land to cropland, although some habitat conversion is still occurring. Indeed, the reported area of unimproved land increased slightly to 31 percent in 1991.[24] This is the most valuable wildlife habitat and includes prairie grasslands, wetlands and woodland.

Prairie grasslands comprise the largest proportion of unimproved land. In 1991, prairie farmers reported 13.8 Mha of unimproved pasture. This represents 25 percent of the prairie landscape, with cultivated areas

[22] Unimproved pasture is occasionally sprayed with pesticide for grasshopper control on the prairies (Ted Weins, personal communication)

[23] Agriculture Census, 1911.

[24] T.W. Weins, Sustaining Canada's wildlife habitat, Prairie Farm Rehabilitation Administration, Draft Manuscript, n.d.

for crops occupying 68 percent, tame pastures six percent, and cities and towns one percent. Wetland areas across the prairie provinces comprise about 37 percent of Canada's wetland habitat area. Precise data on the extent of the conversion of these wetlands to agricultural use are not available, but estimates place losses in the order of 40-60 percent (Ted Weins, personal communication). The area of woodland habitat available to wildlife on prairie farmland declined by more than 80 percent since the 1930s.[25]

As a result of these changes, prairie wildlife populations now depend on one quarter or less of their original habitats. Compared to other ecoregions in Canada, the prairies have a relatively high proportion of birds and terrestrial mammals that are threatened or endangered.[26] In Alberta alone, 73 percent, or 16 of 22 wildlife species that are now considered at serious risk rely on prairie habitats[27] and about 25 percent of 324 vascular plant species considered rare are prairie species.[28]

Concern over the degradation of wildlife habitat and the decline of wildlife numbers has sparked involvement in several initiatives:

1. North American Waterfowl Management Plan (NAWMP)

The North American Waterfowl Management Plan (NAWMP) was signed by Canada and the United States in 1986 in response to concerns about the degradation of wetland habitat together with a sharp decline in waterfowl and shorebird populations across North America.[29] The explicit objective of the 15-year agreement was to restore waterfowl and other migratory bird populations to the levels of the 1970s by securing, enhancing, and managing wetland habitat across North America.

The delivery of NAWMP on the Prairies is co-ordinated by the Prairie Habitat Joint Venture (PHJV). Under the PHJV, agricultural producers are encouraged to set aside agricultural land to retain wetlands for waterfowl habitat and to maintain potholes and native uplands for nesting cover as opposed to putting them into crop production. In Alberta, more than 44,800 ha of large and small wetlands have been protected, more than 16,400 ha of grassland have been restored, and almost 68,000 ha of native grasslands and other habitat have been preserved through the PHJV.[30] Under the program, protected parcels of critical wildlife habitat are subject to either outright purchase or leasing arrangements. Stewardship agreements are also established. These involve landowners or private corporations in discussions with representatives of conservation agencies about appropriate land use activities and methods to enhance or restore the value of their land for wildlife, while continuing to farm the land. Sustainable land uses promoted through the stewardship agreements benefit the entire landscape as well as the individual landowner. For example, preventing erosion on one property can benefit the entire watershed, and enhance general agricultural productivity. Entire communities are becoming involved in these stewardship programs. Enhanced wildlife habitat also attracts tourists, who in turn contribute to the local economy.

---

[25] *Op. cit.*

[26] Environment Canada, A Report on Canada's Progress Towards a National Set of Environmental Indicators, State of the Environment Report No. 91-1, 1991.

[27] Alberta Fish and Wildlife, The Status of Alberta Wildlife, Alberta Forestry, Lands and Wildlife Publication No. I/413, Edmonton, 1991.

[28] G.W. Argus and K.M. Pryer, *Rare Vascular Plants in Canada: Our Natural Heritage* (Ottawa: Canadian Museum of Nature, 1990).

[29] Mexico signed the agreement in 1994 making the NAWMP a truly continental effort.

[30] Agriculture and Agri-Food Canada, *Agriculture in Harmony with Nature: Strategy for Environmentally Sustainable Agriculture and Agri-food Development in Canada* (Ottawa: Minister of Public Works and Services, 1997).

As a result of these efforts there are indications that the goal of NAWMP has essentially been met in recent years for most species of waterfowl in North America's "duck factory" – the Canadian prairies and Northwest Territories and the United States Dakotas. Widespread restoration of habitat under the NAWMP, coupled with two years of exceptional rains that filled the prairie's potholes (i.e. small ponds) have been responsible for a significant resurgence in numbers:

> With tall grass now protecting the nests of land breeders from predators, plenty of water for diving ducks to breed in and lots of food for all, the number of breeding ducks among the 10 most abundant species reached an estimated 35.9 million in the traditional survey area this year, according to the U.S. Fish and Wildlife Service.[31]

The goal of the NAWMP for the survey area is 36 million, which represents the average count of the 1970s. Despite recent successes, biologists and conservationists remain concerned. Legislation proposed in the United States Congress would cut back on programs that restore habitats. If passed, the new law would gradually reduce annual money for the Conservation Reserve Program to $974 million (United States) from $1.8 billion over the next seven years. It would also reduce the upper limit of the number of hectares conserved from 15 million to 10 million. There are concerns that the proposed cuts would more than wipe out recent gains and would constitute a major set back for the ducks and the ecosystems in which they breed.

2. Waterfowl Production Benefits of Fall-Seeded Cereal Grains

Croplands can represent important nesting habitat for waterfowl. About two percent of all mallard nests in the Canadian aspen parkland occur in cropland.[32] Other species such as teal and northern pintail also use croplands at high rates. Nests initiated in croplands, especially early nests, are vulnerable to destruction by periodic farming operations, such as cultivation, seeding and spraying. Given that up to 85 percent of upland habitats in some important waterfowl breeding areas are annually cultivated, strategies to offset this risk are being sought.

Research is being conducted by the Institute for Wetland and Waterfowl Research on the potential benefits of conservation tillage croplands to waterfowl production.[33] No-till winter wheat and fall rye require minimum tillage operations, provide residual cover and are planted in the fall when there is no conflict with the duck nesting period. Preliminary results of the density and success of nests initiated in this fall-seeded cropland are encouraging. Fall-seeded cereal grains have the potential to provide suitable nesting habitat for at least six dabbling duck species. Observed nest density was approximately one nest for every four hectares and as high as one nest for every two hectares. Hatching success was somewhat higher than the normal average of 10 to 15 percent observed in non-cropland habitat. Given the extent of the area involved, the conversion to conservation tillage could make a significant contribution to improved breeding conditions for waterfowl.

3. Flushing Devices in Hay Fields

Upland nesting ducks, particularly late nesting species, such as blue-winged teal (*Anas discors*), gadwall (*A. strepera*), or renesting mallards (*A. platyrhynchos*), often nest in hayfields. Haying operations result in high nest losses and losses of nesting females. A variety of other wildlife species, including ground-

---

[31] W. Stevens, Suddenly, it's ducks unlimited, *Globe and Mail*, 18 November, 1995

[32] Institute for Wetland and Waterfowl Research, PHJV Assessment Program, Procedures Manual (unpublished), Institute for Wetland and Waterfowl Research, Ducks Unlimited Canada, Stonewall, MB, 1995.

[33] J.H. Devries, *Waterfowl Production in Fall-Seeded Cereal Grains in Saskatchewan* (Stonewall, MB: Institute for Wetland and Waterfowl Research, Ducks Unlimited Canada, 1996).

nesting songbirds, a variety of reptiles and amphibians, and small mammals such as mice and voles (family *Cricetidae*), and shrews (family *Soricidae*), also use hayfields for nesting and escape cover. They suffer high mortality during haying.

Flushing devices designed to reduce wildlife mortality during hay mowing were first used in the late 1940s but were considered too heavy and cumbersome as mower equipment evolved. Based on research findings which demonstrate the effectiveness of flushing devices in reducing female duck mortality, Ducks Unlimited have urged the design and development of more effective and convenient flushing technology.[34] Given that all hayfields provide habitat to a variety of species, they recommend that flushing devices should be used in all hayfields. The promotion of flushing device use in target areas of high wetland density and correspondingly high waterfowl pair numbers is the first priority.

4. Environmentally Sustainable Agricultural Policies

A number of agriculture policies with implications detrimental to conservation have been changed (e.g. certain provisions in the Canadian Wheat Board Quota System) or abandoned (e.g. Alberta's On-farm Wetland Drainage Program). In recent years, initiatives such as the Permanent Cover Program (1989-93), offered by the Prairie Farm Rehabilitation Administration (PFRA), have been promoted. This program resulted in the removal of 555,000 ha of marginal prairie agricultural land from annual crop production or summerfallow. Placing these lands under permanent cover reduces the risk of soil degradation and has direct benefits for wildlife.[35]

Trends into the Future

Over the past two decades there has been a net shift to more sustainable agricultural practices. Almost 70 percent of prairie farms are now using at least one erosion control practice to control the environmental stress imposed by traditional monoculture systems (Table 6.2). Changes in cropping and tillage practices are producing significant improvements in soil health.[36] A seven percent reduction in the risk of wind erosion over the past 10 years has been attributed to the reduced use of summerfallow and increased use of conservation tillage and other erosion controls. The risk of water erosion decreased by eleven percent for the same period, in addition to a decrease in the risk of salinity because of the use of salinity controls (e.g. permanent cover, and extended crop rotations).

In addition, the contribution of agriculture to Canada's total greenhouse gas emissions decreased from eight percent in 1990 to about six percent in 1995 as a result of increased sequestering of carbon in soils, together with measures to manage agricultural sources of greenhouse gases.[37] These measures include: decreased soil tillage and area in summerfallow, decreased use of fossil fuels (associated with savings in heavy equipment use), enhanced efficiency of nitrogen fertilizer, improved feeding technology for ruminant livestock, and improved manure handling and storage.

Further improvements are expected as more farmers adopt these methods. In the interim, the news is not all positive. Under 1991 management practices, about 15 percent of prairie cultivated land had soil erosion rates that exceeded the tolerable annual limit.[38] About 4.9 Mha of marginal Prairie land continues

[34] B.K. Calverley, and T. Sankowski, Effectiveness of tractor-mounted flushing devices in reducing accidental mortality of upland-nesting ducks in central Alberta hayfields, Alberta NAWMP Centre, NAWMP-019, Edmonton, 1995.

[35] Prairie Farm Rehabilitation Administration, *A Branch Overview,* Agriculture Canada, Regina, 1992.

[36] D.F. Acton and L.J. Gregorich, *op. cit.*

[37] National Agricultural Environment Commitment, Greenhouse Gases and Agriculture, Fact Sheet, Ottawa, 1995.

[38] Agriculture and Agri-Food Canada, *op. cit.*

to be cultivated annually. And while various regulatory and private actions have had modest success in restoring uncultivated areas that serve as refuges for native species and ecosystems, some farming interests have recently lobbied to curtail this trend.

As a society, we are increasingly aware that sustainability in economic and social terms is linked to ecological sustainability. The case of prairie agriculture suggests that a quiltwork of areas under intensive production, together with preserved and restored wetlands, grasslands, and woodlands, is the practical approach to meeting the twin goals of sustainability and biodiversity conservation. Areas under intensive production should themselves mimic the ecological advantages of the native prairie, and this is likely, in the long run, to enhance rather than restrict productivity from a human standpoint. The case also underlines the fact that it is the complex and interwoven effect of government policy, advocacy by non-governmental interest groups, and the choices and practices of individual rural producers, that will determine outcomes.

Questions

1. What are the goals of sustainable agriculture, and how does it differ from conventional agriculture?

2. This case study demonstrates that sustainable agricultural practices not only make sound economic sense but also have significant benefits for wildlife. Despite this the majority of agricultural producers continue to use conventional agricultural practices. Suggest possible explanations to account for the persistence of these conventional, and often unsustainable, practices.

3. It has been suggested that one of the major difficulties in any attempt to preserve wetland habitats is the limited recognition given to the non-market values of wetlands and waterfowl. Discuss.

Further Reading

Biodiversity Science Assessment Team. 1994. Biodiversity in Canada: A Science Assessment for Environment Canada. Ottawa: Environment Canada.

Gray, J.H. 1996. *Men against the Desert.* Saskatoon and Calgary: Fifth House Publishers, 2nd edition.

Gliessman, S. R., ed. 1990. *Agroecology: Researching the Ecological Basis for Sustainable Agriculture.* New York: Springer-Verlag.

Sopuck, R. D. 1993. Canada's Agricultural and Trade Policies: Implications for Rural Renewal and Biodiversity. Ottawa: National Round Table on the Environment and the Economy.

Hilts, S. G and A. M. Fuller, eds. 1990. *The Guelph Seminars on Sustainable Development Guelph*: University School of Rural Planning and Development, University of Guelph.

Fulton, M. E., Rosaasen, K., and A. Schmitz 1989. *Canadian Agricultural Policy and Prairie Agriculture.* Ottawa: Economic Council of Canada.

Spector, D. 1983. *Agriculture on the Prairies 1870-1940.* Ottawa: National Historic Parks and Sites Branch.

Web Sites

- Winter Wheat Production, U. of Saskatchewan: http://www.usask.ca/agriculture/cropsci/winter_wheat
- Prairie Farm Rehabilitation Administration: http://www.agr.ca\pfra
- Sustainable agriculture: http://www.maritimes.dfo.ca/science/mesd/he/lists/theory/msg00181.html
- Definition of sustainable agriculture: http://www.agrenv.mcgill.ca/EXTENSION/EAP/sustain.htm
- History of Sustainable Agriculture: http://www.agrenv.mcgill.ca/EXTENSION/EAP/history.htm
- Concept of Sustainable Agriculture: http://www.sarep.ucdavis.edu/sarep/concept.html

Audio-Visual Material

- *Agriculture in Crisis: New Approaches to Farming*, National Film Board, 1990 (dir.: Floyd Elliot and Tamara Lynch, 19 min.).
- *Ducks in Danger*, Canadian Broadcasting Corporation, 1985 (51 min.).
- *Prairie Grasslands: Wind Country*, Karvonen Films Ltd. and National Film Board, 1992 (dir.: Joseph Viszmeg, 48 min.).
- *Sustainable Development and the Ecosystem Approach*, General Assembly Post Productions Services for the State for the Environment Reporting of Environment Canada, 1993 (18 min.).

Case Seven
Polar Bears: The Politics of Protection

Focus Concept

In the absence of accurate population estimates, the establishment of an international polar bear conservation regime, and the imposition of quotas on aboriginal communities, may have had more to do with international politics, and the state's old habit of monopolizing management authority, than with ecological necessity.

Introduction

The polar bear (*Ursus maritimus*) "is a charismatic animal: large, powerful, playful, fierce, human-like, *white* (white animals of almost any kind seem to hold a peculiar attraction for people), a survivor of the most vigorous environmental conditions in the world, *and* international in its distribution."[1] In the absence of sound population estimates it is these characteristics that have largely defined the way polar bears have been managed over the past three decades.

Although polar bears have never been classified as endangered, a marked increase in the number of polar bears being killed in the 1950s and 1960s gave rise to world-wide concern that the species might be endangered. Several scientific conferences in the late 1960s and early 1970s called attention to their vulnerability to over-harvesting as a result of low reproductive rates and growing threats to the Arctic ecosystem. Since no reliable scientific data were available on the status of the polar bear population, animal protection organizations chose to highlight the lowest population estimates and called for the designation of the polar bear as an endangered species. Frequently cited estimates of 5,000-10,000 in a context of annual world harvest figures of 1,300 to 1,500 suggested that the polar bear was severely threatened and seemed a strong case for an international moratorium on polar bear taking. Animal rights and welfare groups attracted considerable attention in the media and fuelled public concern over the status of the polar bear. At the same time newly established environmental organizations such as Greenpeace and the International Fund for Animal Welfare (IFAW) were gaining world-wide support as public concern about a number of environmental issues grew; the impact of oil exploration in the Arctic, harvesting of baby seals, leg-hold traps, commercial whaling, etc. These factors combined to create a favorable political environment in which to negotiate the terms of an international regime for the conservation of polar bears.

> On balance, we believe that a sense of crisis, or at least a sense of urgency, at once reported in and generated by the media, created an environment conducive to regime formation and increased the probability of success in the negotiations...[2]

The *International Circumpolar Agreement on the Conservation of Polar Bears*, signed in November 1973, was unique in having the support of all five states with coastlines bordering the Arctic Ocean – Canada, Denmark, Norway, the Soviet Union, and the United States. Despite deep divisions between these members on military and strategic issues, they readily agreed to form a united front on the

---

[1] A. Fikkan, G. Osherenko and A. Arikainen, Polar bears: the importance of simplicity, in O. R. Young and G. Osherenko, eds., *Polar Politics: Creating International Environmental Regimes* (Ithaca and London: Cornell University Press, 1993), 96.

[2] *Op. cit.*, 135.

protection of the polar bear. This involved the coordination of national measures to protect the polar bear, to protect the ecosystems of which the polar bears are a part, and to manage polar bear populations "in accordance with sound conservation practices based on the best available scientific data."[3] The real motivation for their involvement, however, may have had less to do with concern for the ecological status of the polar bear than a perceived opportunity to further international relations: "The superpowers saw in this beloved and apolitical animal an opportunity to lessen international tensions and improve circumpolar and international relations."[4]

The degree of uncertainty in scientific knowledge about polar bear populations, combined with the international profile of the issue, contributed to a regime that has catered to the priorities and interests of the five signatory parties. The possibility that this resulted in overly conservative management that compromised the interests of less powerful stakeholders is explored in the following case study.

Biological Attributes

Polar bears (*Ursus maritimus*)[5] are widely believed to have evolved from brown bears in Siberia during the glacial advance of the mid-Pleistocene period (100,000 to 250,000 years ago). Isolated by glaciers, these brown bears underwent a rapid series of evolutionary adaptations that enabled them to survive in one of the world's most hostile environmental settings. Seven distinct polar bear populations are recognized today: central Siberia, Svalbard Archipelago, Greenland, the Canadian arctic archipelago, northern Alaska, western Alaska and Wrangel Island, and the Hudson Bay-James Bay populations. These groupings have developed as a result of separate ice movement patterns.

Polar bears mate out on the pack ice in April/May. Through a remarkable process of delayed implantation, the fertilized ovum divides a few times and then floats free within the uterus for about six months with its development arrested. Sometime around late September/early October, the embryo implants itself in the uterine wall and resumes its development. Females enter maternity dens by late October and the young, usually two, are born in late December or early January. The cubs are small, weighing about 1 kg at birth, and still have their eyes closed. In most of the Arctic, family groups leave the dens in late March or early April, when the cubs weigh 10-15 kg. Cubs generally remain with their mothers for two-and-a-half years. Consequently, an adult female produces a litter only once in three years.

Male polar bears weigh 500-600 kg and reach maximum size and sexual maturity in eight to ten years. Females are about half the size of males, weigh 200-300 kg and reach maximum size and sexual maturity in four to five years. In the wild natural longevity for males is 20+ years, and for females is 25+ years.

Polar bears feed primarily on ringed seals but also on bearded seals and occasionally on walruses, beluga whales and narwhals. Since polar bears are at the top of the food chain they are valuable indicators of the health of the marine ecosystem.[6] If a polar bear population is healthy then it is likely that other components of that food chain are doing well. If problems arise with polar bear, it may indicate problems elsewhere within the ecosystem. Top predators such as polar bears have been found to develop the highest

---

[3] Article II and Preamble, *International Circumpolar Agreement on the Conservation of Polar Bears*, signed in November 1973.

[4] A. Fikkan, *et al.*, *op. cit.*, 96.

[5] "sea bear"

[6] R.D.P. Eaton and J.P. Farant, The polar bear as a biological indicator of the environmental mercury burden, *Arctic*, 36(3), September 1982, 422-245.

concentrations of organochlorines and heavy metals. This is because they concentrate the contaminants present in the animals and plants they consume.[7]

Distribution and Movements

Polar bears inhabit most of the ice-covered seas of the northern hemisphere. Their preferred habitat is near the edges of leads in the ice or in areas where the ice regularly cracks open because of wind and currents and then refreezes.[8] These areas of open water surrounded by ice are called polynyas. Recurring polynyas, ones that occur in the same place each year, are of great biological importance because wintering or migrating marine mammals and birds can rely on them as breeding and feeding areas.

Several years ago it was thought that all polar bears in the Arctic might be part of one circumpolar population nomadically roaming over the Arctic.[9] Scientists now recognize that polar bears are distributed in many discrete, highly mobile, sub-populations rather than in one continuous population; however, the precise pattern and extent of intermingling across international boundaries is still unclear.[10] Polar bears have also been found to exhibit strong seasonal fidelity to local areas (home ranges).[11] This explains the results from the tagging and recapture of several thousand polar bears (including return of ear tags from animals shot by Inuit and Indian hunters) which indicated a tendency for polar bears to be recaptured close to the original site of capture.[12] Their distribution, seasonal movements, and size of home range are principally determined by two factors: 1) seasonal variations in the presence and distribution of sea ice which provides them with a platform from which to hunt seals; and 2) the distribution, abundance, and availability of the ringed and bearded seals upon which they prey. Thus in places where ice conditions or low food abundance preclude seals, there tend to be fewer polar bear home ranges, and bears may have to move many hundreds of kilometres each year to remain on the ice. Conversely, in areas of high biological productivity where seals are abundant, there will be many overlapping polar bear home ranges. Together these home ranges form what is called, for conservation purposes, a sub-population.

Based on information about seasonal fidelity and natural barriers to movement, recent analyses of mark-recapture and harvest data, twelve sub-populations of polar bears are recognised in Canada (Figure 7.1).[13] A remarkable feature of these sub-populations is their separation in places where there are no apparent natural barriers.[14] For example, bears of northwestern Ontario and James Bay appear to form a separate sub-population from those in western Hudson Bay. This separation is even more surprising in the Beaufort Sea where two sub-populations are maintained despite a lack of apparent landmarks to navigate

---

[7] Animals at each ascending level of the food chain have higher levels of toxic chemicals in their tissues than those below them. This phenomenon is known as bio-accumulation.

[8] I. Stirling, D. Andriashek, and W. Calvert, Habitat preferences of polar bears in the western Canadian Arctic in late winter and spring, *Polar Record*, 29(168), 1993, 13-24.

[9] A. Pederson, *Der Eisbär: Verbreitung Und Lebensweise* (Copenhagen: E. Bruun and Company, 1945).

[10] M. Taylor and J. Lee, Distribution and abundance of Canadian polar bear populations: a management perspective, *Arctic*, 48(2), June 1995, 147-154.

[11] I. Stirling, D. Andriashek, C. Spencer, and A. Derocher, Assessment of the polar bear population in the eastern Beaufort Sea, Final report to the Northern Oil and Gas Action Program (Edmonton: Canadian Wildlife Service, 1988).

[12] Mark-recapture techniques involve the capture of polar bears mainly from helicopters using remote injection equipment. Each animal captured is permanently marked with an individual identification tattoo and ear tags before being released. Some bears are also fitted with radio collars to provide additional information on bear movements.

[13] M. Taylor and J. Lee, *op. cit.*

[14] The boundaries of these sub-populations have been determined by a satellite radio-collaring program (R. Bethke, M. Taylor, S. Amstrup, and F. Messier, Population delineation using satellite collar data, *Ecological Applications*, 6(1), 1996, 311-317).

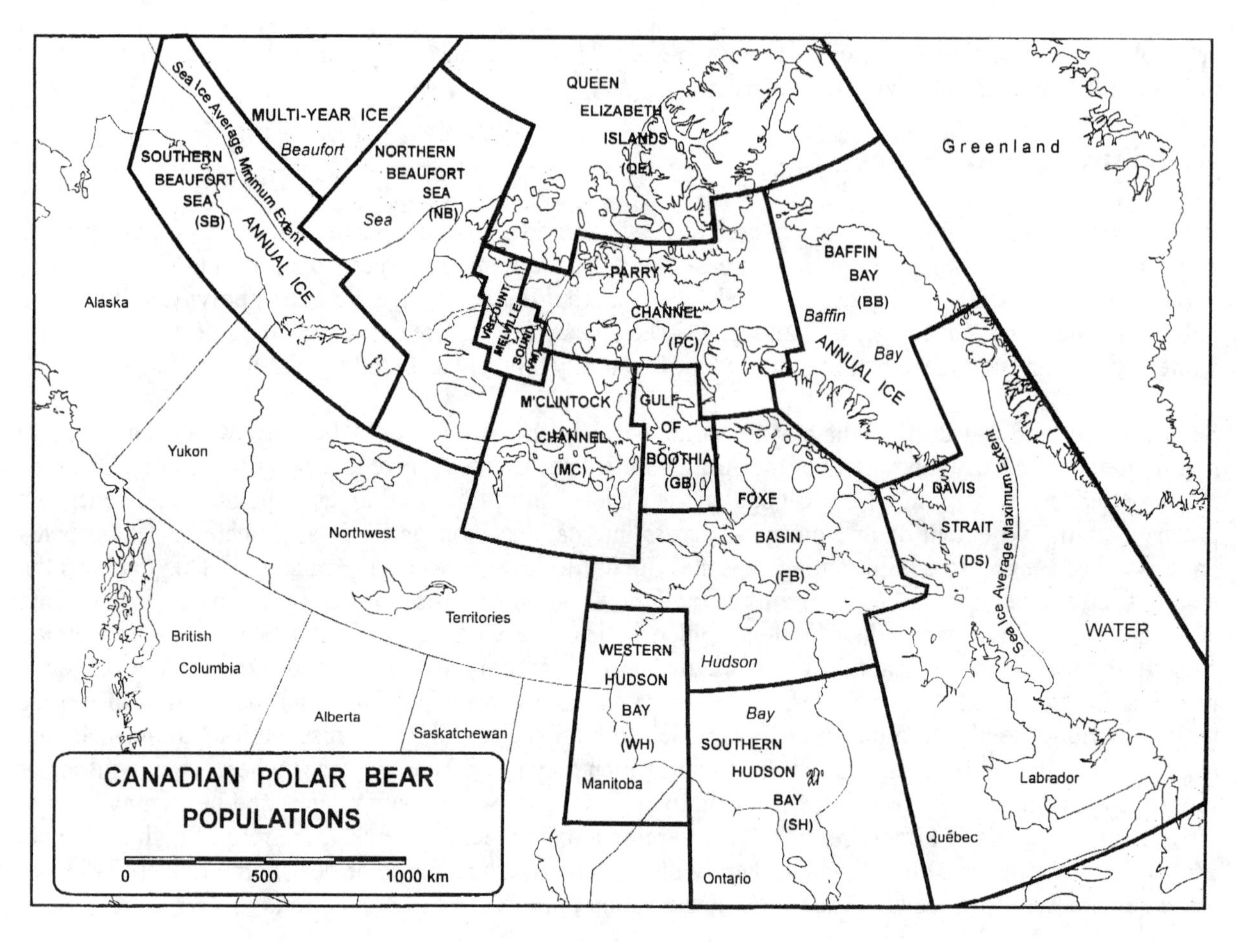

Source: Department of Resources, Wildlife and Economic Development, Northwest Territories, with permission.

Figure 7.1. Canadian Polar Bear Sub-populations

by, and in the presence of a gigantic clockwise circulation pattern of the pack ice, known as the Beaufort Gyre. Bears have learned to compensate for this movement in order to stay in the same area.

Population Size

The world population of polar bears is unknown. The first preliminary estimates by both Canadian and former Soviet Union scientists were as low as 5,000 to 10,000;[15] Uspenski and Shilnikov estimated a world population of 10,000 animals based on a sample of only 58 bears seen on aerial survey flights made over the sea ice north of the former Soviet Union in 1962, 1966, and 1968.[16] These figures could neither be confirmed nor rejected, but if they were even approximate, there were concerns that the future of polar bears was in some doubt.[17] General agreement was reached at the Eighth Working Meeting of the IUCN

[15] S.M. Uspenski and F.B. Chernyavski, Maternity home of polar bears, *Priroda,* 4, 1965, 81-86.

[16] S.M. Uspenski and V.I. Shilnikov, Distribution and numbers of polar bears in the Arctic according to data of aerial ice surveys, *The Polar Bear and its Conservation in the Soviet Arctic* (Leningrad: Hydrometeorological Publishing House, 1969), 89-102.

[17] I. Stirling, Research and management of polar bears *Ursus maritimus*, *Polar Record,* 23(143), 1986, 167-176.

Polar Bear Specialists Group, that the world population was greater than 20,000 and could be as high as 40,000.[18]

Part of the problem with these estimates has been the limited information available on polar bear populations in Russia. Most of the research in Russia has been aimed at ecology and behavior. No mark and recapture studies have been conducted and population estimates are based almost exclusively on extensive aerial surveys.[19] Stirling calculated that the world population of polar bears could total 19,175; however, given the inaccuracy and incompleteness of the various population estimates, he suggested that the total population could be as large as 40,000.[20] A more conservative estimate, based on a review of the available data, was put forward in 1994 by the Eleventh Working Meeting of the IUCN Polar Bear Specialists Group. It suggests that the total world population of polar bears is between 21,370 and 28,270.[21]

According to Stirling, these global estimates, regardless of their accuracy or inaccuracy, are not particularly useful for conservation or management purposes. He suggests that information on the size of separate sub-populations and their ranges is more important.[22] Twelve distinct sub-populations of polar bear have been identified in Canada based on an analysis of the movements of marked and recaptured/killed polar bears (Figure 7.1).[23] Population estimates were obtained using various mark-recapture models. The density of polar bears was then estimated by dividing the population estimates by the average area of sea ice that constitutes polar bear habitat in April. The total area of available habitat was 3.1 million km^2. Density estimates of polar bears ranged between 1.1 and 10.4 bears per 1000 km^2 with a weighted mean of 4.1 bears per 1000 km^2. The sum of polar bear estimates population within or shared with Canada is approximately 12,700. Extrapolation of the mean density observed in Canada to the global area of available bear habitat suggested a total world population of 28,000 polar bears, a figure consistent with the recent PBSG estimate.

Estimating the size of a population where individuals do not concentrate at any one time of year, and in which the probability of sighting is always less than unity, presents unique problems for wildlife biologists. Mark-recapture techniques are considered to be the most cost-effective approach in such situations. This is because in addition to providing population estimates, mark-recapture techniques provide information on bear movements, rate of growth, reproduction and survival.[24] The problem, however, with the population estimates is that the intensity of sampling varies geographically so that there are probably more bears than the mark and recapture estimates suggest. Furthermore, because polar bears are not easy to see on ice or snow and are spread out over large, relatively inaccessible areas at low densities, sample sizes are often too low to be valid regardless of the technique applied. Thus there are many areas where polar bears are present but not represented in the population estimate. The accuracy of mark-recapture estimates has also been limited by non-random capture of animals and poor estimates of adult survival.[25] Additional problems arise because mark-recapture estimates involve the application of closed population models, which assume no net movements in or out of an area and no births or deaths

---

[18] Eighth Workshop Meeting of the IUCN Polar Bear Specialists Group, Oslo, Norway, 1981.
[19] According to Stirling, *op. cit.*, 79, aerial surveys are ineffective for counting polar bears.
[20] I. Stirling, *Polar Bears* (Michigan: The University of Michigan Press, 1988).
[21] Polar Bear Specialist Group, Polar bears, in Ø. Wiig. E. Born and G.W. Garner, eds., *Proceedings of the 11th Working Meeting of the IUCN/SSC Polar Bear Specialist Group 1993*, Occasional Papers of the IUCN Species Survival Commission (SSC, 1994).
[22] I. Stirling, *op. cit.*, 1988.
[23] M. Taylor and J. Lee, *op. cit.*
[24] D.P. De Master, M.C.S. Kingsley and I. Stirling, A multiple mark and recapture estimate applied to polar bears, *Canadian Journal of Zoology*, 58, 1980, 633-638.
[25] M. Taylor and J. Lee, *op. cit.*

over the sampling period. The result of these difficulties is an extremely costly and often unreliable approach to the estimation of population size.

> Research on the population ecology of polar bears is both time-consuming and very expensive. Although we have estimates of the sizes of several sub-populations, we are far from having an initial assessment of all areas. Even in some areas where research has been conducted, the estimates of population size can only be called educated guesses.[26]

As a result of these difficulties there has been a tendency to assume that the total population of polar bears is small, especially in relation to the total number of bears killed annually. In the absence of scientific certainty, biologists have tended to underestimate: "usually we work with what we feel are reasonable yet conservative estimates. In this way, the gamble is that if we make a mistake, it will be on the bears' side."[27] As estimates are refined, however, it is clear that there are many times the number of polar bears projected by the early studies.

International Conservation Status of the Polar Bear

Through the 1950s and 1960s a marked worldwide increase was recorded in the numbers of polar bears being killed; estimates rose from about 600 a year in the 1950s to a high of 1,500 annually in the late 1960s.[28] This led to concerns among conservationists that polar bears might be in danger of extinction and there were calls for polar bears to be given complete protection. Public outrage over the way in which some polar bears were being legally hunted also led to a call for international rules and cooperation.

In Alaska until 1950, only Inuit hunted polar bears and they rarely took more than 120 per year. The use of aircraft for hunting polar bears began in the late 1940s and continued until it was stopped by the State of Alaska in 1972. During this period guided aircraft hunting was responsible for a threefold increase in trophy harvesting of polar bears. State regulations introduced in 1961 provided a preference for subsistence hunters and protected cubs and females with cubs. Trophy hunters were allowed to hunt only during late winter and spring, and an annual limit of three bears per person was imposed. In 1972 the *Marine Mammal Protection Act* (MMPA) vested management with the federal government. Under the Act, all polar bear hunting by non-native people ceased.[29] The Act removed restrictions on the taking of cubs and females with cubs and the mandatory reporting of kills. The only restriction on the Native take is that it must be done in a non-wasteful manner; there are no restrictions on the number, sex, or age of animals taken by Natives hunters. The federal government may, however, restrict the take if the polar bear population is determined to be depleted. Today the polar bear population in Alaska is considered healthy and on the increase, although there are some concerns about the vulnerability of certain age and sex classes of bears. According to a 1986 estimate, there are about 3,000-5,000 polar bears in Alaska; Alaskan Natives take approximately 130 polar bears per year for subsistence use.[30]

In Norway, extensive hunting by trappers, sealers, trophy hunters, and local Norwegians had, by the end of the 1960s, placed polar bears in danger of extinction on the Svalbard Archipelago. The practice of shooting of polar bears from Norwegian tour ships received much international media attention. In response the Norwegian government banned the use of snowmobiles and aircraft for polar bear hunting in

[26] I Stirling, *op. cit.*, 1988, 77.
[27] *Op. cit.*, 141
[28] F. Bruemmer, *World of the Polar Bear* (Toronto: Key Porter Books, 1989), 101.
[29] The importation of polar bear hides was also prohibited under the Act.
[30] The polar bear and the walrus (http://www.teelfamily.com/activities/polarbear)

1967. The use of "set-guns" was prohibited in 1970[31] – this practice involved indiscriminate killing and wounding of bears, regardless of age or sex class. Hunting of cubs and females with cubs was also banned in 1970, and quotas were introduced. A complete ban on the taking of polar bears, except to protect life and property, was imposed in 1973.

Denmark has no polar bears; however, it exercises jurisdiction in Greenland. Although never endangered, it was believed that the polar bear population in Greenland had declined from 1920 until the 1960s, when it was stabilized. Several regulations and restrictions on hunting were introduced in the 1950s, including the protection of cubs and females throughout the year, and all bears during the summer. Hunting from aircraft, snowmobiles, and motorboats was also prohibited and a variety of restrictions on rifles and ammunition were introduced. Commercial hunting of polar bears has been eliminated in more recent years, and only subsistence hunting by Native residents is allowed. No quotas have been imposed but hunting is restricted by the requirement that hunters use only traditional means (which does not include motorized vehicles). Complete protection for resident polar bears is provided in the world's largest national park – Northeast Greenland National Park established in 1973 – which extends over almost one third of Greenland.

The Soviet Union prohibited all hunting of polar bears in 1956. The only reported takes since then have been a few cubs captured each year for public display.

Polar Bear Populations in Canada

Canada has the largest population of polar bears in the world.[32] This population has never been endangered,[33] although historical accounts suggest that populations were reduced locally wherever sealing and whaling industries were located, and as a result of increased harvesting by indigenous peoples for the hide trade.

A decline in Canada's polar bear population was officially cited in 1935 when the Federal Government instituted the first closed season; hunting was limited to a season between 1 October and the following 31 May. The Northwest Territories (NWT) provided additional protection in 1949 when hunting rights were restricted to indigenous peoples. The increasing value of polar bear hides in the 1950s and 1960s, combined with an increase in the use of snowmobiles, led to a rapid rise in the number of polar bears taken. The annual kill in the NWT fluctuated between 325 and 550 from 1953 to 1964, and rose steadily to an unprecedented level of 726 bears in 1967.[34] No accurate means of accessing the impact of these harvests on the population existed at that time – estimates for the period are speculative.

Nonetheless, hunting regulations were introduced in 1967 which imposed a quota system on each native settlement that had access to polar bears. Quotas limiting the annual harvest were based on a crude estimate of the local polar bear population and the average of the previous three years' harvest. The entire quota for the first season (1967-68) was 375 bears. By 1981, that figure had risen to 597 through

---

[31] A set-gun is normally a rifle with a bait attached to the trigger by a cord, the gun is set in such a way that an animal pulling at the bait will have the shot fired into it.

[32] I. Stirling and N. Lunn, Some information about polar bear research in western Hudson Bay, http://www.brandonu.ca/CNSC/polar.htm, n.d.

[33] Polar bears are currently listed by the Committee for International Trade of Endangered Species (CITES) in Appendix II and are classed as "Vulnerable" by the Committee on the Status of Endangered Wildlife in Canada (COSEWIC).

[34] R. Schweinsburg, A brief history of polar bear management in the NWT, *Northwest Territories Wildlife Notes*, 2, 1981, 1-5; cited in P. Prestrud and I. Stirling, The International Polar Bear Agreement and the current status of polar bear conservation, *Aquatic Mammals*, 20(3), 1994, 113.

increments granted, according to Urquhart and Schweinsburg either in response to political pressures, or research results, or both.[35] Quota controls are maintained by issuing tags to the Area Wildlife Officer who transfers them to the local Hunters and Trappers Association, where they are allotted to members of the community. Quota increases are usually in the form of special red tags, which can only be used after the regular quota is filled. These tags can be revoked if there are indications of a population decline. From 1970, part of the quota could be sold to sport hunters under conditions designed to benefit local communities. Under this provision, a small number of the tags, allocated to the annual quota, are used by Inuk hunters to guide non-resident sport hunters. This allows for a greater economic return than does the sale of the hide alone.

In addition to the quota system, hunting regulations introduced in 1967 include: a closed season to protect pregnant females and family groups; a size restriction (minimum length 150 cm) to protect cubs; and protection of family groups. Reporting is mandatory. Voluntary compliance with quotas, seasons, and other regulations is fostered inasmuch as local people are favoured to reap the economic benefits, through the legal sale of hides and management of a limited trophy guiding industry.

International Agreements

The first international meeting to discuss the conservation of polar bears was held in Fairbanks, Alaska in 1965. Representatives from all five polar bear countries (Canada, Denmark/Greenland, Norway, the former Soviet Union, and the United States) attended. Two crucial points became clear – a significant number of polar bears cross international boundaries, and sizable bear populations use areas outside the jurisdiction of any government. Thus, even if individual nations adopted significant conservation measures, there was a need for an international regime.

Agreement was reached in Fairbanks on the basic principles and goals of an international regime: polar bears are an international circumpolar resource; each country should take whatever steps are necessary to conserve the polar bear until the results of more precise research findings can be applied; cubs and females accompanied by cubs should be protected throughout the year; each nation should, to the best of its ability, conduct a research program on polar bears within its territory; and international exchange of research and management information is essential.

In the years following the first international meeting in 1965, the size of the polar bear kill continued to increase. Many countries decided not to wait for the results of long-term research studies or for the negotiation of an international agreement. By the early 1970s, several countries had changed their domestic laws in accordance with the consensus among them that the polar bear needed fuller protection, especially from such practices as hunting from planes and the use of set-guns. In addition, the Polar Bear Specialist Group (PBSG) held regular meetings from 1968 onward to coordinate scientific research and exchange of data. Thus by the time the formal *International Agreement on the Conservation of Polar Bears and their Habitat* was signed in Oslo, Norway in November 1973, a fledgling international regime had already been established.

The Agreement entered into force in May 1973 for an initial period of five years, after which time it became permanent. The Agreement was remarkable from a political standpoint, in bringing together the five Arctic Nations to negotiate and reach mutual agreement on a unique circumpolar concern. The Agreement is also significant from a biological perspective in that it was one of the first international regimes based on ecological principles:

---

[35] D.R. Urquhart and R.E. Schweinsburg, Polar bear: life history and known distribution of polar bear in the Northwest Territories up to 1981 (Yellowknife: Northwest Territories Department of Renewable Resources, 1984).

> Each Contracting Party shall take appropriate action to protect the ecosystems of which polar bears are a part, with special attention to habitat components such as denning and feeding sites and migration patterns, and shall manage polar bear populations in accordance with sound conservation practices based on the best scientific data.[36]

The Agreement establishing the regime was also a first in that it begins by prohibiting all hunting, capturing, and killing of a species, then specifies particular exceptions to the general prohibition. These include *bona fide* scientific purposes; conservation; prevention of serious disturbance to the management of other resources; hunting by local people exercising their traditional rights; and protection of life and property. Thus, the parties agreed to a prohibition regime which would be applied equally to everyone, with exemptions to achieve "fairness" by accommodating the priorities and interests of each state.[37] Canada and Denmark sought to protect subsistence hunting; Canada needed to have troublesome bears captured and transported away from Churchill, Manitoba; and the United States wanted to create a large sanctuary in which no hunting would occur. Norway's need to cull polar bear stocks in cases where bears are interfering with the seal harvest was also accommodated, as was the Soviet Union's desire for a broad prohibition on hunting. The approach taken was to achieve consensus by accommodating the priorities and interests of each state:[38]

> Although the polar bear is a finite resource shared by several states, the negotiators in this case were able to avoid questions of allocation by applying a general (and therefore equitable) prohibition on hunting, with limited exemptions and restricted methods of harvest.[39]

A third significant aspect of the regime is that it extends the prohibition on the taking of bears to all areas frequented by polar bears.[40] This covers international waters in addition to the sovereign territory or jurisdiction of the signatory parties. The terms of the Agreement are not, however, legally enforceable in any country and there is no infrastructure to oversee compliance. The Agreement avoids unnecessary intrusion into the management systems of the five signatory states by maintaining their authority to manage polar bears in coastal waters and on land. According to Stirling, this decentralized approach is an effective and simple form of compliance which encourages countries to honor the terms of the Agreement.

Fikkan *et al.* provide an interesting discussion on the power play and political dynamics which framed the terms of the Agreement. In their view discrepancies between the greater and lesser powers served as obstacles to bilateral talks, but encouraged the formation of multilateral agreements between nations with shared stocks of polar bears. Mutual veto power, coupled with shared interests in protecting the bears, created ideal conditions for integrative bargaining in which there were perceived joint gains through coordinated action. Scientists and managers who formed the PBSG envisaged that they would obtain better information on polar bear populations, and thus management would be enhanced.

---

[36] Anon., *Agreement on the Conservation of Polar Bears and their Habitat*, Article II, 1973.

[37] A. Fikkan, *et al.*, *op. cit.*

[38] While the negotiation of a general prohibition on the taking of polar bears was politically expedient, some critical ecological issues were not addressed by the Agreement. A later resolution appended to the Agreement was required to prohibit the taking of cubs or females with cubs and to restrict hunting in denning areas during certain periods. A second resolution established an international system of identifying hides to effectively control the trafficking of illegal hides.

[39] A. Fikkan, *et al.*, *op. cit.*, 131.

[40] *Op. cit.*

Case Seven

Native Hunting Rights

Polar bears have long been extremely important to circumpolar indigenous peoples, both economically and culturally. Among Canadian Inuit, for example, polar bears are highly prized for several reasons. In addition to the substantial value of the meat and hide, there are various ointments and cures derived from the bear. It has, moreover, outstanding religious and ritual significance. The monetary or commercial value of the bear, originating with the sale of hides, now extends to Inuit control of guiding operations for sport hunters, and to the Korean demand for the polar bear gall bladders, which bring several thousand dollars apiece.[41]

Canadian polar bear populations are not considered to be in danger, but hunting is regulated by a quota system, under which some 600 to 700 bears are taken annually by native hunters, mainly Inuit.[42] It is ironic that in Canada in 1967, when the federal government began imposing quotas on native communities, there was no compelling evidence that polar bear populations were in fact threatened.[43] Indeed, estimates of polar bear numbers rose steadily over the next dozen years as scientific information gradually improved. Perhaps the best face that could be put on this dramatic intervention by the state in the aboriginal rights and livelihoods of native hunters was that government authorities felt obliged to "err on the side of caution." Clearly the scientific community was aware of the deficiencies in its own knowledge base. Jonkel, for example, observed frankly in 1970 that "much has been written about Polar Bears in scientific and popular form, but it is mostly guesswork and conjecture, if compared with research on other species."[44] Nonetheless, great political mileage was made of the ill-founded population under-estimates of the 1950s and 1960s, which generated a sense of crisis, and seemed to justify the imposition of restrictions.

Once the issue became international, Canada's strong interest in protecting its autonomy in managing the largest polar bear population in the world ensured its participation. At the same time, Canada acted, according to Fikkan *et al.*, to protect the traditional rights of its indigenous people to hunt polar bears, and this became a central issue in the negotiation of the international polar bear regime.[45] Although the Soviet Union had wanted a total ban on hunting of polar bears, all parties agreed that indigenous people should be allowed to continue traditional hunting, within safe, scientifically established limits. The language used to describe this Native exemption from the ban on hunting was the subject of much debate. Objections to the terms "indigenous" or "aboriginal" peoples resulted in the reference to the "traditional rights of local people" in the final draft of the Agreement.

Since 1982, aboriginal rights in Canada have had explicit constitutional recognition, and Supreme Court rulings have tended to recognize the primacy and ability of aboriginal people to hunt without restriction, except as required for conservation, as part of these rights. This has created a climate in which quotas, as well as seasonal and other restrictions, are unworkable without aboriginal consent, except, in theory,

---

[41] The total monetary value of polar bears to the NWT is about 1.7 million dollars annually. This includes the polar bear fur harvest which currently brings between 500,000 to 600,000 dollars per year into the Arctic communities. Outfitting and guiding non-Inuit polar bear hunters average another 660,000 dollars. A conservative annual replacement cost for polar bear meat is about 150,000 dollars, while the tourism value of the polar bear is 200,000 dollars per year (J. Lee and M. Taylor, Aspects of the polar bear harvest in the Northwest Territories, Canada, *International Conference on Bear Research and Management*, 9(1), 1994, 237-243).

[42] The annual average for polar bears harvested or killed to defend property or life was 692 in 1986/87 – 1990/01 (M. Taylor and J. Lee, *op. cit.*, 147).

[43] A. Fikkan, *et al.*, *op. cit.*, 135.

[44] C.J. Jonkel, Some comments on polar bear management, *Biological Conservation*, 2(2), 1970, 117.

[45] Despite Canada's efforts to protect traditional Native harvesting, traditional hunting received a major set back in 1972 as a result of the federal U.S. *Marine Mammal Protection Act* (MMPA) that prohibited the import of all polar bear products.

where iron-clad scientific evidence permits a government Minister responsible to determine unequivocally that a harvested population is at risk.

Hunting Quotas and Maximum Sustainable Yield

In Canada, efforts to regulate native hunters focused primarily on limiting harvests (through a system of quotas) and closed seasons. Scientific wildlife management normally involves the calculation of safe harvest limits through mathematical models of the maximum sustainable yield (MSY) of a population; these, however, are only as good as one's estimates of the total population. Given the difficulties of estimating polar bear population (see above), it is questionable whether these "top down" models are as useful as native hunters' reliance on "ground up" observations of *trends* in population, which provide a basis for decisions to increase or decrease harvests without needing to know total numbers. In fact, estimates of total populations for northern species have proven notoriously unreliable, susceptible to political manipulation by state managers, and frequently at odds with what local hunters know.[46]

The PBSG has found that the paucity of information on age-specific survival and recruitment rates makes simulation of sub-populations unreliable for setting quotas on populations of polar bear. An alternative approach to setting quotas is based on the sustainable harvest of adult (i.e. non-cub) females from the population, the sex ratio of the harvest, and an estimate of population size. This approach has the advantage of being easy to understand and explain, uses the more reliable data on harvest and population size, and assumes that most bears are similar enough in their life histories that a sustainable yield estimate of female bears will apply across populations.[47]

Formerly, biologists thought that a polar or grizzly bear population could be harvested safely at a maximum level of about 5 percent if the harvest was spread evenly across all the age and sex classes, excluding females with cubs of the year. Mitch Taylor, polar bear biologist for the NWT, suggests that the safe harvesting level is probably closer to about 1.5 percent of adult females before the population begins to decline:

> Polar bears have an extremely low capability of sustaining a harvest. Only 1.5 percent of the total population of adult females can be harvested each year ...if you take more than that, the population declines. The problem is that the decline is so gradual that it's virtually undetectable until the bears have been reduced to the point that you have a numerical effect on the number taken, and then there's a quick drop off.[48]

More recent studies indicate that under optimal conditions the sustainable yield of adult female polar bears is typically < 1.6 percent of the total population.[49]

The proportion of females in the harvest is assumed to be relatively constant from year to year, so that as long as the quota is not reached, the number of female bears taken will be less than the estimated sustainable yield. To be accurate, this approach requires that reliable estimates of population size are

---

[46] See, for example, M.M.R. Freeman, Graphs and gaffs: a cautionary tale in the common-property resources debate, in F. Berkes, ed., *Common Property Resources: Ecology and Community-Based Sustainable Development* (London: Belhaven Press, 1989), 92-109.

[47] W. Calvert, I. Stirling, M. Taylor, J. Lee, G.B. Kolenosky, S. Kearney, M. Crete, B. Smith, and S. Luttich, Polar bear management in Canada 1985-87, *Proceedings of the Tenth Working Meeting*, IUCN/SSC, 1988

[48] T. Domico, *Bears of the World* (New York/Oxford: Facts on File, 1988).

[49] M. Taylor, D.P. Messier, F.L. Bunnell and R.E. Schweinsburg, Modeling the sustainable harvest of female bears, *Journal of Wildlife Management*, 51(4), 1987, 811-820.

available. Where this is not the case, the PBSG suggested using minimum estimates of population size, so that the quota would be conservative.

This approach underscores the survival of adult females as the most important age and sex class for the maintenance of population viability. It also suggests that a relatively large number of males can be harvested from a population before the population begins to decline. Indeed while Eberhardt acknowledges that bear populations with delayed maturity and associated lower reproductive rates[50] need to be managed with caution, he suggests that some "polar bear populations might support larger harvests than are often assumed in the literature."[51]

Discussion

It is argued by some specialists that the exclusion of hunters from critical breeding areas and the protection of females and cubs is likely to make an even more significant contribution to the sustainable harvest of a population than the imposition of quotas. Stirling is critical of the *Marine Mammal Protection Act* for not regulating these variables, and views its adoption by Alaska as a backward step in the management of its polar bears.[52] The Act reduced the total kill of polar bears in Alaska by prohibiting the non-native take of polar bears; however, it provided no restrictions on subsistence hunting. Since there was no longer a closed season or protection for bears in dens or females with cubs, a significant proportion of the kill has been concentrated on the most valuable portion of the population, the reproductive females.

This concern, however, begs the question of whether such native harvests are beyond sustainable limits. While harvesting females may not be consistent with maximising a population's growth rate, it can nonetheless be practiced within sustainable limits. From the perspective of local hunters, cultural values, technological considerations, or specific ecosystemic circumstances might override the scientific modeller's implicit goal of maximising population growth, especially if the local population of bears is not depressed. Indigenous hunters of large mammals in various cultures worldwide have practiced the preferential harvesting of the pregnant females of certain species over long periods, without undermining those populations. The key issue, in such cases, is whether the harvests have been kept within sustainable limits by the overall effect of local management practices.

Denmark has opted for an approach different from the quota system. The indigenous population of Greenland has been harvesting 100 to 150 polar bears annually for many years. Although no population studies exist, there are no apparent signs of overharvest and no quotas have been imposed. Instead, various regulations and restrictions on the taking of females and family groups have been imposed in addition to a requirement that subsistence hunting be limited to traditional methods. Fikkan *et al.* suggest that the taking of bears mostly by dog sleds, which are slow and with a shorter range, has prevented an uncontrolled increase in the harvest.[53] Stirling has used this same argument to explain the sustainable harvest of bears by aboriginal peoples over the past thousands of years: "unlimited hunting with stone age technology did not threaten the survival of the polar bear."[54]

Such arguments reflect a common bias on the part of western scientists, which is to underestimate the role of traditional/local knowledge and resource management practices. More powerful technological means

---

[50] Usually a consequence of limited food supplies.

[51] L.L. Eberhardt, Survival rates required to sustain bear populations, *Journal of Wildlife Management*, 54(4), 1990, 590.

[52] I. Stirling, *op. cit.*, 1986.

[53] A. Fikkan, *et al., op. cit.*

[54] I, Stirling, *op. cit.*, 1988, 187.

need not lead to overharvesting, where local institutions of knowledge, management, and social control respond to declines in resource availability by self-limiting harvests.[55] State intervention may actually be detrimental, from a conservation point of view, if it interferes with the workability and local legitimacy of these institutions.

All social institutions evolve, and generally do so more effectively if continuity can be maintained between past adaptations and contemporary innovations. It should not simply be assumed that new technologies, growing populations, or even the harvesting of animals for commercial purposes, are circumstances beyond the ability of local knowledge and management practices to accommodate. The primary source of conflict in negotiating the *International Polar Bear Agreement* was not indigenous harvesting *per se,* but whether Inuit and Indian communities in Canada would be permitted to continue to use part of their quota for hunting by non-indigenous sportsmen with Native guides.[56] Restricting hunters to the harvest of animals for "non-commercial" subsistence purposes only, like confining them to the use of "traditional" technologies, seems to say to indigenous hunting societies that they must lose their rights if they wish to modernize.

There were few restrictions on the numbers of polar bears shot in the Arctic before the establishment of quotas in 1968 and the cessation of sport hunting in Alaska in 1972; the unstated assumption, according to Stirling, was that the population could support the harvest.[57] The call for international protection was mainly in response to concerns about growth in trophy hunting and commercial hunting. Yet this culminated in imposition of restrictions on indigenous hunters who, to all indications, were not depleting polar bear populations; by means of a state management system that has by no means demonstrated, even today, its ability to act as effectively as local institutions. The latter are under-researched and poorly understood by non-Natives.

Prestrud and Stirling cite "equivocal" evidence on polar bear numbers from around the circumpolar north to suggest that "following a period of overharvesting in the 1960s, the population increased up to about 1978 after which it has been relatively stable."[58] Population growth in the wake of regulatory action by the state, however, is not in itself evidence that populations were previously "depressed" below optimum levels. When the rate of harvests and other mortality is less than a population's rate of recruitment, it will tend to rise to carrying capacity. But it is well known that the maximum biological productivity of prey species typically occurs when the population is maintained well below its maximum carrying capacity. For many species, maximum reproductive rates are reached at an "optimal" population level of around 50% of carrying capacity. Aboriginal hunters understand that reproductive rates decline as a population becomes too numerous – in the parlance of scientific ecology, when numbers escalate beyond that optimum level at which sustainable yields are maximized. Given the state of scientific knowledge of polar bear populations prior to the imposition of quotas, it would be guesswork to say whether these populations were ever below optimal levels in the Canadian Arctic. Contemporary science, however, in

---

[55] see, for example, H. Feit, Waswanipi Cree management of land and wildlife: Cree ethno-ecology revisited, in B. Cox, ed., *Native People, Native Lands: Canadian Indians, Inuit and Metis* (Ottawa: Carleton University Press, 1992), 75-91.

[56] Canada interprets traditional rights to permit the locally guided sport-hunt to continue. Animal rights groups denounce trophy hunting as "barbaric," and it is also often decried by aboriginal hunters, many of whom see it as wasteful and disrespectful to the animal. On the other hand, native guiding increases the chances that maximum use will be made of the parts of any bear killed. From a conservation point of view, because not all guided hunters succeed in killing a polar bear, and because tags cannot be reused, not all the bears in an allowable quota are taken. In this regard, guiding activities not only provide a valuable source of revenue to northern communities, but may result in a reduced harvest.

[57] I. Stirling, *op. cit.*, 1986.

[58] P. Prestrud and I. Stirling, *op. cit.,* 118.

combination with Native local knowledge, could address the question of whether more recent increases have put polar bear populations into sub-optimal ranges closer to their maximum carrying capacity.

The Prospects for Co-Management

In sum, we are left with several questions that are only partly resolved or resolvable at this point in time. Are restrictions on the indigenous hunt necessary? If so, to what extent is self-regulation adequate to the need? Are quota systems useful? Given the uncertainty about the size of the different sub-populations and disagreements among biologists over the maximum sustainable yield (MSY), how can quotas be sensitized to local circumstances? Are reduced hunting seasons and prohibitions on the taking of females and adult groups workable or acceptable as alternatives to quotas? How can the right and ability of local people to judge these issues for themselves be worked into management arrangements?

The Beaufort Sea polar bear sub-population is shared internationally between the United States and Canada, and hunted by both the Inupiat of Alaska and the Inuvialuit of Canada. In Canada there are strict management regulations determining quotas, closed seasons and the protection of females with cubs and bears in dens. In Alaska, since the passage of the MMPA in 1972, aboriginal people are allowed to kill polar bears for subsistence purposes without restriction. Prestrud and Stirling, despite their disapproval of this regulatory approach, state that annual harvests by indigenous people there have stabilized at 100-150 animals.[59] This suggests, at the very least, that the hypothesis of Native self-management of harvests be taken seriously. Research on the effect of these contrasting approaches in Alaska and Canada would provide important information about the efficacy of "self-management" versus "state management."

The political representative bodies of aboriginal groups are now playing a major role in the protection and management of polar bears and other species. As part of land claims agreements, these bodies participate, with Territorial and Provincial governments, in the co-management of wildlife. There is also significant lateral cooperation between aboriginal groups. In the Northwest Territories, management agreements between user communities, using both indigenous and scientific knowledge, and witnessed by the Minister of the Department of Renewable Resources, are negotiated to ensure coordinated harvests within sustainable limits.[60] This type of arrangement now also crosses international boundaries. Local interest in the Beaufort polar bear population led to an unofficial agreement between the Alaskan and Canadian aboriginal user groups. The *Management Agreement for Polar Bears in the Southern Beaufort Sea* was formally signed in Inuvik in January 1988 by the Inuvialuit Game Council of Inuvik and the North Slope Borough Fish and Game Management Committee of Barrow, Alaska. Among other measures, this cooperative management agreement calls for establishing harvest limits, protecting females and cubs, protecting denning bears, collecting critical information from harvested polar bears, and preserving important polar bear habitat. These measures are foreign neither to indigenous approaches, nor to state managers.

There is, in fact, considerable potential for a symbiosis of state management and self-management in co-management arrangements. "Ground-up" and "top-down" approaches may be more powerful in unison than either would be in isolation. Indigenous knowledge and self-management may well play the more effective lead role on the local scene, while the state may be required to take the lead in assessing and regulating external pressures that are beyond the means of native communities to control.[61]

---

[59] *Op. cit.*, 114.
[60] *Op. cit.*, 116.
[61] Prestrud and Stirling provide a discussion of several of these pressures, including "long range transportation of hydrocarbons, offshore exploration for and production of hydrocarbons, radioactivity from nuclear dumping, climatic warming, and increased disturbance and harassment resulting from increased development in the Arctic in general," *op. cit.*, 113.

A vision of polar bear management for the 21st century involves decentralized, culturally autonomous systems, linked cooperatively to the initiatives of central governments, who in turn can build on their history of international collaboration. Indigenous peoples themselves, through such bodies as the Inuit Circumpolar Conference and the Arctic Council, are directly involved in transnational knowledge-building and policy-making. The time for hierarchical notions of central government monopoly on management policy, and for naïve assumptions about the inherent superiority of scientific information and models, has passed.

Questions

1. What are the management implications of the uncertainties and limitations inherent in scientific knowledge about highly dispersed populations of solitary animals like polar bears?

2. Discuss the ways in which local knowledge and scientific research can complement one another in wildlife management. What factors inhibit cooperation of distinct knowledge traditions?

3. Discuss arguments for and against the imposition of restrictions on the hunting technology used by Native people.

4. Why do some animals hold a particular fascination for humans? How does this influence the level of protection we afford them?

Further Reading

Amstrup, S.C. and D.P. DeMaster 1988. Polar bear, *Ursus maritimus*. In J.W. Lentfer, ed., *Selected Marine Mammals of Alaska: Species Accounts with Research and Management Recommendations*. Washington, D.C.: Marine Mammal Commission, 39-56.

Bruemmer, F. 1989. *World of the Polar Bear*. Toronto: Key Porter Books.

Derocher, A.E. and I. Stirling. 1991. Population dynamics of polar bears in western Hudson Bay. In D.R. McCullough and R.H. Barrett, eds., *Wildlife 2001: Populations*, London/NewYork: Elsevier, 1150-9.

Domico, T. 1988. *Bears of the World*. New York/Oxford: Facts on File.

Lee, J. and M. Taylor. 1994. Aspects of the polar bear harvest in the Northwest Territories, Canada. *International Conference on Bear Research and Management* 9(1):237-243.

Prestrud, P. and I. Stirling. 1994. The *International Polar Bear Agreement* and the current status of polar bear conservation. *Aquatic Mammals* 20(3):113-124.

Stirling, I. 1988. *Polar Bears*. Michigan: The University of Michigan Press.

Web Sites

- Agreements Related to Polar Bears: http://maple.nis.net/~bearwork/polars1/02intagr.html
- Churchill Northern Studies Centre: http://www.brandonu.ca/CNSC/welcome.htm
- International Wildlife Education and Conservation: http://www.iwec.org/bear.htm
- Polar Bears: http://www.nature-net.com/bears/polar.html
- Polar Bear: http://www.teelfamily.com/activities/polarbear
- Polar Bear Collection: http://maple.nis.net/~bearwork/polars1/13seawrl/pbbigindex.html
- Polar Bear Home Page: http://eagle2.eaglenet.com/jcurley/jfc/pbear.html
- Research in Western Hudson Bay: http://www.brandonu.ca/CNSC/polar.htm

Audio-Visual Material

- *Polar bears*. National Film Board of Canada, 1984 (dir.: Floyd Elliot)

Case Eight
Banff National Park: Defining Ecological Integrity

Focus Concept

Developing a world class tourist destination without compromising the ecological integrity of a national park.

Introduction

Banff National Park is the birthplace of Canada's national park system and the second national park established in North America.[1] In just over a century since the first reserve was set aside, the Park has become a major tourist destination, currently attracting more than 5 million people annually and generating about half a billion dollars annually in tourist revenue. Development projects within the Park have been undertaken to keep pace with the rapid increase in visitor numbers (Figure 8.1). Facilities and services include three ski resorts, one golf course, a major resort town (Banff), a resort hamlet (Lake Louise), over 3,600 hotel rooms, over 2,500 campsites, 125 restaurants and 220 retail outlets. Some 8,000 people reside permanently within the Park's boundaries, while a large transient population provides services to the millions of visitors who visit the park each year. A national transportation corridor, which includes the Canadian Pacific Railway (CPR) and the Trans-Canada Highway (TCH), bisects the Park.

Banff National Park is also prime habitat for wildlife and a central component of the larger Rocky Mountain Ecosystem, which stretches from Wyoming to the Yukon. The Park was designated as a UNESCO World Heritage Site in 1983 in recognition of its exceptional natural beauty and its importance as a habitat for threatened species.[2] Development is, however, disrupting wildlife habitat, and threatening vital wildlife movement corridors for large mammals such as wolves, bears and cougars.

The issue of how to balance development and wildlife preservation in Banff National Park has become increasingly urgent in recent years. A recent report on the likely future prospects of the Park predicts that if trends of the past 50 years continue unabated, 19 million visitors per year could visit the Park by 2020. Even with a more modest growth rate of 3 percent, the annual number of visitors could exceed 10 million by the same date.[3]

In an attempt to find a solution to this dilemma, the Minister of Canadian Heritage and Deputy Prime Minister, Sheila Copps, commissioned an independent task force to conduct a two year study of the cumulative environmental effects of development and human use on the park. The findings of the Study, released in October 1996, concluded that over-development and overuse threaten Banff National Park's ecological integrity and status as a national park:

> this growth in visitor numbers and development threatens the mountain environment. If allowed to continue, it will cause serious and irreversible harm to Banff National Park's

---

[1] The first national park in North America was established at Yellowstone in 1872.

[2] Designation as a World Heritage Site requires that four "natural" criteria be met including Evolutionary History, On-going Geological Processes, Exceptional Natural Beauty, and Habitats for Threatened Species. Banff National Park fulfils all four criteria.

[3] J. Green, C. Pacas, S. Bayley and L. Cornwell, eds., *Banff-Bow Valley Futures Outlook Project: A Cumulative Effects Assessment and Futures Outlook of the Banff Bow Valley,* Prepared for the Banff Bow Valley Study (Ottawa: Department of Canadian Heritage, 1996).

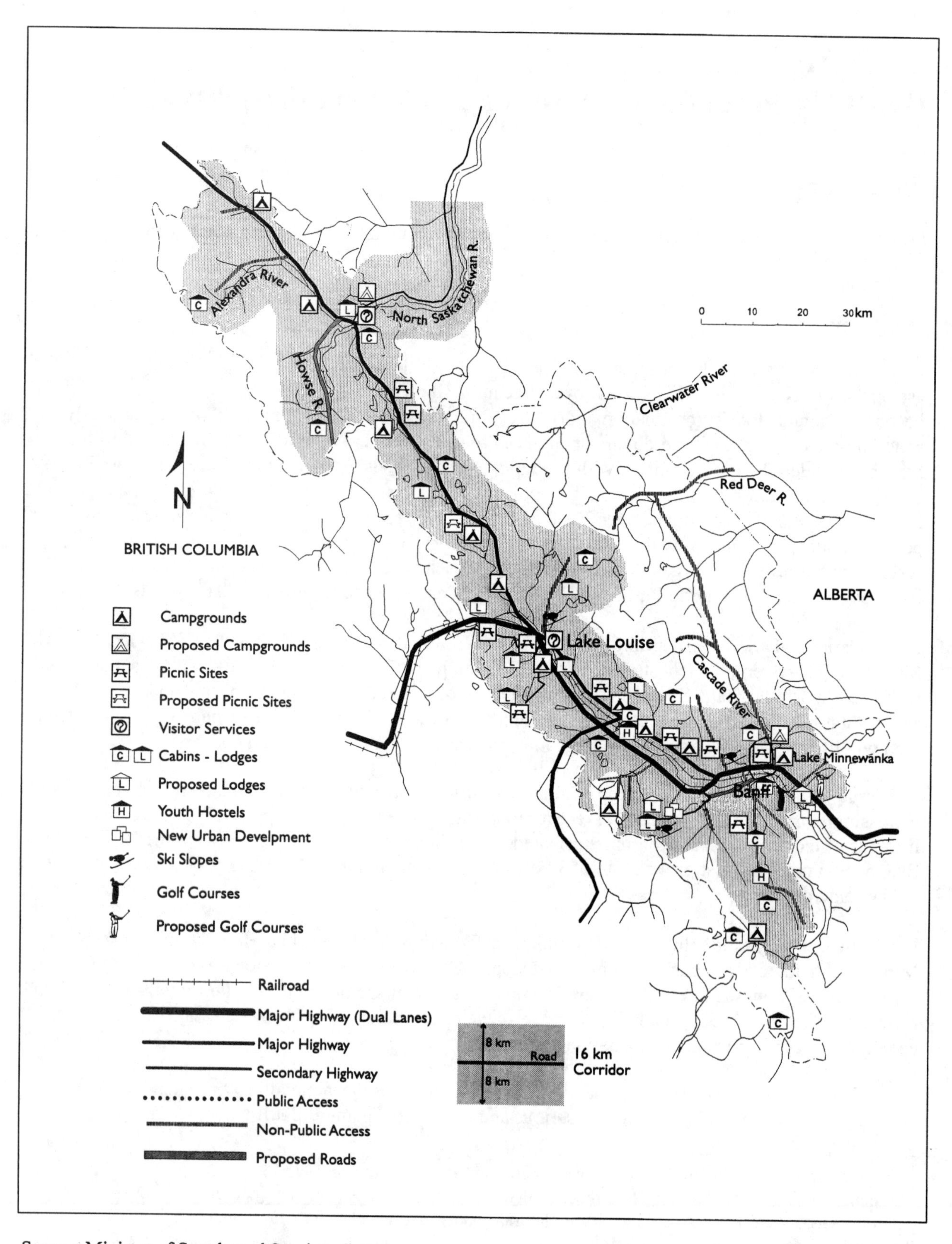

Source: Ministry of Supply and Services Canada

Figure 8.1. Facility Development in Banff National Park, 1992.

ecological integrity and its value as a national park. Impairing the ecological integrity and natural beauty of the Park will also weaken its attraction as a tourist destination, and the associated contribution to the local, regional and national economies.[4]

This case study describes some of the major ecological impacts of development and human use within Banff National Park. The principle of ecological integrity provides the framework for the discussion.

History of Banff National Park

The origin of Banff National Park dates back to 1871 when Prime Minister John A. MacDonald promised residents of British Columbia that a railway would be built linking the eastern and western regions of the country if they joined Confederation. In November 1883, one month after the CPR reached the site of Banff, three railway workers discovered the natural hot springs in a hillside just above the Bow Valley. Two years later, MacDonald reserved an area of 10 square miles[5] around the hot springs "as a public park and pleasure grounds for the benefit, advantage, and enjoyment of the people of Canada."[6]

The park was expanded to 260 square miles in 1887, the same year that Parliament introduced the *Rocky Mountain Parks Act*. Modelled on the United States legislation that created Yellowstone National Park, the Act stressed public enjoyment and recreation. The Park was expanded again in 1902 to include the area around Lake Louise. This secured a large habitat that would provide for the protection of wildlife species; however, resource development, such as logging and mining, was permitted and hunting of elk, bear and other game was allowed within Park boundaries. Conservation was a growing interest, however, and major policy debates at the time centered on the merits of resource development versus natural resource conservation. As a result, the *National Parks Act* of 1911 excluded new mining and commercial lumbering and introduced the principle of appropriate use: "there will be no business except such as is absolutely necessary for the recreation of the people."[7] In the same year, however, lobbying efforts of ranchers and resource industries resulted in a reduction in the Park's area by about 60 percent. Although a counter campaign by conservationists resulted in the restoration of some parts in 1917, the Park's reduced area was the political price for strict rules on appropriate use.[8]

Tourism changed fundamentally during the mid-1900s as roads and highways pushed through the mountains, and motels, cabins and campgrounds were established to meet the needs of the tourists. Following the completion of the Banff section of the TCH in the 1960s, developers began to entertain thoughts of Banff as an all-season tourist destination. During the 1960s and 1970s, as Park facilities were refurbished and reoriented to meet the demands of a new year-round market, public concern about national parks increased dramatically. This coincided with the emergence of an active environmental movement in Canada and came to a head with the bid for the 1972 Olympics at the Lake Louise Ski Resort. The level of public controversy over the Games forced the federal government to convene public hearings and ultimately led to the rejection of the proposal. The credibility of Parks Canada was severely undermined by this episode, but it lent legitimacy to the role and influence of environmental groups, and opened up public debate over the issue of appropriate use in national parks.

---

[4] R. Page, S. Bayley, J.D. Cook, J.E. Green, and J.R.B. Ritchie, *Banff-Bow Valley: At the Crossroads*, Summary Report of the Banff-Bow Valley Task Force to the Hon. Sheila Copps (Ottawa: Minister of Canadian Heritage, October, 1996) 4.

[5] A standard unit of land measurement in the Canadian west is one square mile, equivalent to 259 hectares.

[6] The *Rocky Mountain Parks Act,* 1887.

[7] As cited in P. Dearden, and R. Rollins, *Parks and Protected Areas in Canada: Planning and Management* (Toronto: Oxford University Press, 1993).

[8] Hydro-electric development was allowed to continue, and was later expanded within the Park under the *War Measures Act.*

During the last thirty years, park management has shifted between three different federal ministries – Indian and Northern Affairs (1966-79), Environment Canada (1979-1993), and Canadian Heritage (1993 to the present). In addition to adapting themselves to differences in the corporate cultures of these ministries, park managers have been caught between the opposing aspirations of environmental groups and the business community, who compete for center stage in public and political arenas. Tourism is the fastest growing industry in Alberta and Banff is the province's premier tourist destination. On the other side, environmental groups such as the Bow Valley Naturalists and Canadian Parks and Wilderness Society, continue to grow in strength and influence. Parks Canada has tended to avoid controversy and to take the path of least resistance by allowing both sides to proceed with their plans and expectations, until inescapable contradictions erupt in conflict. This has resulted in development decisions that frequently cannot be reconciled with Parks Canada's policies (see below).

The Banff Bow Valley Task Force was appointed when tensions threatened to get out of hand:

> It is our belief, after carefully weighing the evidence we have heard, that current trends, if allowed to continue, will lead to the destruction of the conditions in the Banff Bow Valley that are required for a national park ... We believe Banff National Park is clearly at a crossroads and changes must come quickly if the Park is to survive.[9]

Thus, the 110-year history of Banff National Park traces the complex interplay between the economic development goals and ecological sensibilities that have shaped the Park. The current crisis is the legacy of a long and tangled history of *ad hoc* arrangements, compromises and concessions:

> when you look at what you see in Banff National Park, it is not a result of decisions made necessarily today or even yesterday or ten years ago. We're talking about a park that has evolved over a hundred years when our understanding of issues and priorities may have been very different.[10]

Banff-Bow Valley Study

The Banff-Bow Valley Task Force was commissioned in March 1994 by the Ministry of Canadian Heritage in response to growing concerns for the ecological integrity of Banff National Park. Its job, at a cost of $2 million, was to assess the cumulative environmental effects of development and use in the Park, and to make recommendations concerning the long-term management of the area that would maintain ecological integrity while allowing appropriate levels of development and continued access for visitors.[11]

An independent Task Force comprised of five people from the academic and private sectors with expertise in ecological sciences, public policy and management was created. Public involvement was a critical component of the Task Force's approach. A Round Table, with representatives from fourteen sectors with interests in the Park met and formulated a "vision" to guide the management of the Park.[12]

---

[9] R. Page, *et al.*, *op. cit.*, 18.
[10] Charlie Zinken, superintendent of Banff National Park as quoted in D. Suzuki, National Parks: forever wild? *The Nature of Things*, CBC, 1994/95.
[11] R. Page, *et al.*, *op. cit.*
[12] *Op. cit.*

Recommendations from the Round Table were subsequently incorporated into the Banff National Park Management Plan, which was tabled at Parliament in April 1997.[13]

Ecological Integrity

During the past three decades, national park policies have experienced a progressive increase in emphasis on environmental protection. The importance of conservation was formally endorsed in the 1964 Parks Canada Policy Statement – "national parks are dedicated to the people of Canada for their benefit, education, and enjoyment" while remaining "unimpaired for the enjoyment of future generations."

This trend in policy was enshrined in legislation with the 1988 amendments to the *National Parks Act*. Maintaining ecological integrity through the protection of natural resources became the first priority in zoning and visitor use management. Ecological integrity was defined as "the minimization of human impact on natural processes of ecological change" and "protecting intact ecosystems".

The most recent Parks Canada policy statement accords ecosystems "the highest degree of protection of natural environments essentially unaltered by human activity."[14] It stipulates that activities which "threaten the integrity of parks ecosystems will not be permitted"; only those activities that maintain or enhance the "ecological and commemorative integrity" of a park are allowed.[15]

The principle of ecological integrity, as defined in the *National Parks Act* and the 1994 policy statement, underwrites the Banff-Bow Valley Study, and is voiced as the guiding principle for all aspects of acquiring, managing and administering the Park. Problems arise, however, in the application of this principle.

Firstly, Parks Canada claims to recognize humans as part of the ecosystem – indeed this would seem to be implicit in their stated objective to protect nature *for* human enjoyment. Yet, paradoxically, the definition of ecological integrity is tied to the protection of so-called "natural ecosystems" from human interference. But what is a "natural ecosystem" in a context of a human history that began perhaps 11,000 years ago, and in an environment regulated for thousands of years through fire management, not to mention other aboriginal peoples' technologies? Contemporary species in Banff have *never* existed independent of major interactions with humans. The notion of the isolation of natural ecosystems (i.e. as unmodified by human activity) is no less problematic in the more recent context – where is an ecosystem to be found that is unmodified by human action over the past 110 years of Banff's history as a national park? What kinds and degrees of human engagement are to be accepted before an ecosystem is considered too modified to be "natural"? Are two human communities, three commercial ski facilities, a 27-hole golf course, a four-lane divided highway and a national rail corridor consistent with "ecological integrity," as contemplated by the Act and the Policy?

The second problem concerns the inconsistent and contradictory application of the principle of ecological integrity. The rationale for the Banff-Bow Valley Study was that the ecological integrity of the Park was jeopardized by human development. Yet a number of major proposals and projects were exempted from the moratorium on development while the Study was being conducted: nine additional holes of golf, 250 new rooms for the Banff Springs Hotel, the expansion of the Sunshine Ski Resort, the construction of a

[13] Parks Canada, *Banff National Park Management Plan*, Canadian Heritage (Ottawa: Minister of Public Works and Government Services, 1997).
[14] Parks Canada, *Guiding Principles and Operational Policies,* Canadian Heritage (Ottawa: Minister of Supply and Services Canada, 1994), section 3.1.1.
[15] *Op. cit.*, section 3.1.2.

convention center and additional staff housing at Château Lake Louise, and the widening of a stretch of the TCH. To what extent do such political concessions make nonsense of the policy and the process?

Nature's Balance

Ecosystems within Banff National Park have co-evolved with a variety of natural disturbances that maintain a wide diversity of vegetation types and wildlife habitat:

> Natural fires, as well as fires set by aboriginal people, were perhaps the most important influence on the montane and subalpine areas of the valley. Flooding along the Bow River and its tributaries is essential for healthy riparian communities. On the outwash or alluvial fans of creeks entering the valley bottoms, turbulent water flows and shifting rocky debris create new habitat for trembling aspens and the diverse communities they support. Avalanches clear areas of trees and shrubs, opening them up for new growth that is essential food for wildlife. Insect infestations and disease may affect some forest stands and wildlife, but over the long term, contribute to the ongoing renewal of the valley ecosystems.[16]

These ecosystems have been exposed to a variety of different disturbances since the occupation of the Park by white settlers in the 1880s. Developments throughout the 20th century in the form of towns, cabins, trails, and other facilities eliminated or altered several natural communities. Problems with air pollution, sewage discharges, solid waste, and demands on potable water grew as the numbers of residents and park visitors increased. Sensitive wildlife have tended to avoid areas of high human activity, thereby alienating additional habitat[17]. The suppression of fire has lead to a gradual aging of forests and a further loss of important wildlife habitat. Although hunting no longer occurs within the Park, animals dangerous to humans are frequently shot or relocated outside the park.[18]

Transportation corridors through the Park, including the TCH and other highways as well as the CPR, fragment the landscape and block the movement of wildlife and colonization of plants. Landscape fragmentation has resulted in habitat patchiness and a decline in habitat effectiveness for grizzly bears and wolves.[19] Collisions with highway vehicles are a major cause of death for elk, deer, moose, wolf and coyote. The railway has modified water flows and altered important wetland habitat. Because it runs through prime wildlife habitat, it also represents a major hazard to wildlife. Several species use the line as a movement corridor, especially in winter when the snow is deep, and many mortalities result. Chemical and grain spills, unless cleaned up immediately, compound the hazard. The present site of the airstrip and adjacent facilities also restrains or prevents wildlife movement through a significant wildlife corridor. Noise and pollution from highway, rail and air traffic are additional problems.

Some of these effects are an unavoidable consequence of human use and residency within the Park; however, a number of changes have been implemented in recent decades to reduce the impacts of humans on the park.[20] These include: improved garbage management; an end to fish stocking; fencing of the twinned sections of the TCH; construction of underpasses and overpasses to maintain habitat

[16] R. Page *et al., op. cit.*, 99.

[17] Habitat alienation refers to the temporary or long-term avoidance of an area of suitable habitat by wildlife as a result of sensory disturbances from human activities and facilities.

[18] R. Page *et al., op. cit.*

[19] *Op. cit.*

[20] *Op. cit.*

connectivity;[21] closure of backcountry roads; temporary closure of some areas to protect sensitive wildlife; reclamation of disturbed sites; restoration of several creeks; removal of fencing around the golf course, buffalo paddocks and public corrals to reduce obstructions for wildlife travel; and participation in various cooperative programs with other jurisdictions.

Despite these measures, the Task Force has documented consistent evidence that the ecosystems of the Banff-Bow Valley and surrounding region "are continuing to be detrimentally affected by previous and current management practices and human development inside and outside the Park."[22]

The Cumulative Effects Assessment Study for the Banff-Bow Valley Study identified the following important environmental concerns:[23]

- Landscape fragmentation due to human activity and facilities;
- Loss of habitat connectivity between major areas of protected habitat as a result of human development and use;
- Loss of aquatic and riparian habitat associated with dams, stream channelization and water regulation;
- The effect of dams on the movement, diversity and viability of fish and aquatic organisms;
- Mortality of fish and wildlife caused by humans;
- The effects of fire suppression and water regulation on vegetation successional patterns;
- Loss of the montane habitats due to human development and fire control;
- Altered predator-prey relationships;
- Wildlife-human conflicts;
- The effect of human activities on water quality; and
- The introduction of non-native plants and fish.

The following section elaborates on some of the major impacts of development and human use on the Montane Ecoregion, where most of the human activities within the Park are concentrated.

The Montane Ecosystem

The Montane Ecoregion is found at the heart of Banff National Park, within the narrow valley floor of the Bow River. Although this area accounts for only 147 km^2, it represents a unique and disproportionately important ecological area. Frequent chinook winds, low snow accumulation, and warm winter temperatures support rich, diverse vegetation and provide habitats for a wide variety of mammals, amphibians and insects, along with over 180 species of birds. This ecoregion also provides prime feeding grounds and an important migration corridor for large carnivores, ungulates and songbirds.

The problem, however, is that most human activities within the Park also occur within this limited area. Thirty-seven percent of the Montane Ecoregion in the park is occupied by facilities related to tourism development. This development, together with other human activities, is resulting in the alteration, fragmentation and loss of natural habitats and is having a significant impact on wildlife populations, including the grizzly bear, wolf, cougar, lynx, wolverine, otter and moose:

---

[21] Fencing has reduced wildlife mortality but has caused problems for wildlife movement as most carnivores are unwilling to use the culvert style underpasses. Two 50 m overpasses have recently been constructed on the TCH and early indications suggest that these reduce wildlife mortality related to road collisions.

[22] R. Page *et al., op. cit.*, 100.

[23] J. Green, *et al., op. cit.*

> The Bow Valley isn't just the most fragile part of the park; it's also the most populated and visited part of the Park. Montane – the fertile land on the bottom of the Bow Valley – may only make up 6.3 percent of the park but it's at the very centre of the park's fragile ecosystem. And it is this unique and sensitive land that represents the prime feeding corridor for park wildlife.[24]

Vegetation

Vegetation plays an essential role in the ecological integrity of the Bow Valley.[25] The composition, structure and spatial patterns of vegetation are greatly influenced by natural disturbances, such as fire, flood, avalanche, landslide and herbivory (grazing and browsing). Humans have altered these disturbances, for example through fire suppression, and have thus altered the vegetation of the area.

The predominant vegetation trend in the Valley is to more closed forest vegetation and to spruce and spruce-fir forests. This has been paralleled by a gradual decline in aspen forests related to the suppression of the fires that, in the past, regenerated them. Fires due to lightning and aboriginal burning practices were integral to ecosystemic processes in the area for thousands of years. Fire prevention and suppression, focusing especially on the Montane Ecoregion, have produced a tenfold decrease in fires since 1930. In addition to the decline in aspen forests, these changes have led to a decrease in post-fire herb and low shrub communities as well as young conifer (pine, spruce, Douglas fir) forests. These losses in floral diversity impact on other species, which depend on these vegetation communities for habitat.

A number of other factors are contributing to the decline in aspen stands:

> The [aspen] trees are nearing their maximum life span and have low vigor due to a combination of ungulate browsing, antler rubbing, insects and diseases, and road salt. Stands are composed predominantly of older trees, the number of younger root suckers is low, and few, if any, escape browsing long enough to develop into saplings or trees.[26]

Elk browse on young aspen clones, and chew the bark of older clones, weakening them. Successful regeneration of aspen will depend upon both the restoration of fire as a periodic disturbance, and the reduction of ungulate browsing.

The occurrence of non-native plants can also have detrimental effects on the integrity of ecosystems in the Bow Valley. The presence of these species is related to human activity in two respects: firstly most have been introduced by humans; and secondly, non-native plants tend to occupy human-disturbed sites almost exclusively. There are currently 77 known species of non-native plants in the Park, all of which occur in the Bow Valley. These include spotted knapweed, Canada thistle, oxeye daisy, tall buttercup, blueweed, leafy spurge, scentless chamomile and tansy.[27]

Large Mammals

Grizzly bears move freely across jurisdictional boundaries in the Central Rockies Ecosystem. The current population of grizzlies in the Park, although estimated at between 60 and 80 bears, is scientifically

---

[24] Deputy Prime Minister and Minister of Canadian Heritage, Hon. Sheila Copps, speaking notes on the release of the Banff-Bow Valley Task Force Report, October 7, 1996.
[25] P. Achuff, I. Pengally and J. Wierschowski, Banff-Bow Valley Study: vegetation module progress report, in *Summary of Presentations: Ecological Outlook Project*, March 1996.
[26] *Op. cit.*, 2
[27] *Op. cit.*

unknown[28]. Demographic studies are currently being conducted on their regional population status by the Eastern Slopes Grizzly Bear Research Project, and a parallel Western Slopes Project, from which population trends and numerical estimates are not yet available. A regional level management plan involving an interagency, multi-stakeholder group is urgently needed.

Management of the grizzly bear population within the Park tends to focus on the problem of bear-human conflict, with resolution of such conflict frequently involving grizzly bear mortalities/removals from the Park. Between 1971 and 1995, 73 grizzlies are known to have died in the park, 52 of which were either destroyed or removed in the interests of public safety. Ninety percent of the grizzlies die close to developed areas, and 56 percent of those lost were females, since females with curious cubs are most likely to run into trouble with people.

Meanwhile, habitat quality for grizzly bears in the Park remains unacceptably low. This problem is related partly to the suppression of natural fires (grizzlies have adapted to forage in the succession environments created by wildfire) and partly to habitat fragmentation (resulting, for example, from the TCH).

The Montane Ecoregion is home to about 40 gray wolves. The presence of these wolves is threatened, however, and their long-term persistence is considered unlikely.[29] Although the Montane Ecoregion comprises the most important wolf habitat in the Central Rocky Mountains, habitat in the Valley is rated at less than 50 percent of its effectiveness under pristine conditions. In addition to various other restrictions imposed by human activity, the town of Banff disrupts east-west wolf movement, while the TCH impedes their north-south movement. Underpasses in the twinned portion of the Highway have had variable success, more often acting as "filters" to animal movement – some species, or age-sex classes within species, will use them while others (including wolves) will not. Travel patterns have been further altered by human activities that modify snow cover – skiing, snowmobiling and ploughing. The result is an uneven distribution of wolves within the valley, which in turn has affected the distribution and abundance of elk – a prey species upon which wolves depend.

Elk are the most abundant ungulate in the Bow Valley, and although they have inhabited the valley for about 10,500 years, their current numbers are unprecedented in history.[30] This is partly related to the presence of few predators, which has allowed them to multiply rapidly. Their remarkable adaptation to living with humans has also resulted in a change in their distribution within the valley, particularly in the last decade. Elk have increased in the central zone, where an estimated 400 elk now live in the town's immediate environs. Their numbers have remained stable in the eastern zone, and have decreased in the western zone. Fencing and wildlife crossings have reduced elk mortalities along the highway within the central and eastern zones, while collisions between elk and vehicles remain common in the western section of the valley where the highway has been partially twinned and fenced.

Bow Valley elk have complex seasonal movements and are partial migrants; in general, mature males migrate seasonally from the valley while mature females remain year round. Seasonal migrations expand the range of the elk to jurisdictions outside the valley and play an important role in gene flow. Elk are primarily grazers, their diet consisting of leafy vegetation in summer and cured grasses and woody

---

[28] S. Herrero, Preliminary conclusions and recommendations for the Banff Bow Valley Study: grizzly bear, in *Summary of Presentations: Ecological Outlook Project*, March 1996.
[29] Anon., Effects of human activities on gray wolves in the Bow River Valley of Banff National Park, in *Summary of Presentations: Ecological Outlook Project*, March 1996.
[30] J. Woods, L. Cornwell, R. Kunelius, P. Pacquet and J. Weirschowski, The Bow Valley Study: the elk model progress report, in *Summary of Presentations: Ecological Outlook Project*, March 1996.

browse in winter. There are growing concerns that the high densities of elk in some areas are having a negative impact on aspen regeneration through browsing, as described earlier.

Changes in elk numbers and concentrations are also impacting on the wolf population.[31] Elk are a primary food source for the wolves, contributing over 80 percent of the biomass consumed by wolves. However, wolf-killed elk are relatively rare in the central zone where the elk are concentrated. This apparent anomaly seems to reflect the intolerance of wolves to human activity compared to the high tolerance to human activity of elk, resulting in substantial modifications to predator-prey relationships in the valley.

The success of the elk may also be linked to the decline of moose, once one of the dominant ungulates, but now largely absent. Moose die on the roads and railway in addition to experiencing natural cycles of population change. However, their decline in recent decades is also related to the increased prevalence and intensity of the giant liver fluke parasite within the valley. Elk are carriers of the parasite and although they occasionally die from it, in general the elk have a high recruitment rate and relatively low natural mortality rate. Moose on the other hand have a low tolerance of this parasite load and have suffered a significant decline in numbers as a result.

Thus large elk populations do not necessarily translate into secure populations of the predators normally dependent upon elk, in contexts of intensive human activity. Finding solutions to such issues as the growing elk population will require more than an understanding of ecosystemic relationships. It will require assigning priorities and values to different components within the park. Some people say parks are for wildlife; others believe that a nature reserve is for wildlife; but a park is for people *and* wildlife.

Task Force Conclusions and Management Response

Upon the release of the Bow Valley Task Force findings on October 7, 1996, the Minister responsible, Sheila Copps, endorsed environmental stewardship and ecological integrity at Banff. She also announced the government's intention to act immediately on some of the Task Force's recommendations (Table 8.1). The proposed actions included the establishment of clear limits to growth. No new parcels of land will be made available for commercial development in the Park, and the population of Banff is to be capped below 10,000, prohibiting its rise to the status of a city. Efforts to restore a critically impacted wildlife movement corridor include the closure and rehabilitation of the airstrip, the bison paddock, the horse corrals and the cadet camp. More wildlife overpasses will be added and efforts to restore aquatic biodiversity in the Park will be improved. Quotas and reservation systems will be implemented on certain trails. The Minister also announced that Canadian Pacific had withdrawn its plans to expand the Banff Springs Golf course and its hotels, and agreed to focus its efforts on improving the visitor experience rather than increasing hotel capacity.

Later in the same month, Prime Minister Jean Chrétien declared before an international audience at the World Conservation Congress in Montreal: "Banff is important to Canadians and to our government. We are determined to protect the ecological integrity of Banff for Canadians and the citizens of the world – forever." And not long thereafter, a Supreme Court of Canada decision in February 1997 ordered a comprehensive review of the cumulative environmental effects of the Sunshine Ski Resort expansion project, notwithstanding its exempt status under the federal moratorium on development in the Park. This represents the first full environmental review of a development in Banff National Park and the first in any Canadian park – a major victory for environmentalist involvement in parks policy.

[31] Anon., *op. cit.*

A new comprehensive park management plan for Banff was tabled in the House of Commons in April 1997, incorporating the main elements of the Task Force's recommendations (Table 8.1). The future of the park is to be guided by the following "core vision":

> Banff National Park reveals the majesty and wildness of the Rocky Mountains. It is a symbol of Canada, a place of great beauty, where nature is able to flourish and evolve. People from around the world participate in the life of the park, finding inspiration, enjoyment, livelihoods and understanding. Through their wisdom and foresight in protecting this small part of the planet, Canadians demonstrate leadership in forging healthy relationships between people and nature. Banff National Park, is above all else a place of wonder, where the richness of life is respected and celebrated.[32]

Table 8.1. Conclusions and Recommendations of the Banff Bow Valley Task Force

1.	Banff National Park suffers from inconsistent application of the National Parks Act and Parks Canada Policy, *ad hoc* decision making, and weak political will.
2.	The Park's ecological integrity has been, and continues to be, increasingly compromised.
3.	Current rates of growth in visitor numbers and development, if allowed to continue, will cause serious, and irreversible, harm to Banff National Park's ecological integrity.
4.	More effective methods of managing and limiting human use in the Park are required.
5.	To maintain natural landscapes and processes, disturbance such as fire and flooding must be restored to appropriate levels in the Park.
6.	Existing anomalies in the Park, such as the TCH, CPR, and the Minnewanka dam, must be updated in accordance with the most advanced science, and ecological and engineering practices.
7.	The role of tourism must be refocused and upgraded to reflect the values of the Park and contribute to the achievement of ecological integrity.
8.	Current growth in the number of residents, and in the infrastructure they require, is inconsistent with the principles of a national park.
9.	New forms of broader based public involvement and shared decision-making are needed.
10.	Visitors must be better informed about the Park's natural and cultural heritage, the role of protected areas and the challenges that the Park will face in the third millennium.
11.	A comprehensive revision of the Banff National Park Management Plan is required.
12.	Funding must be made available to meet the requirements for maintaining ecological integrity and visitor management.

A key component of the Park's vision will be the adoption of an integrated approach to decision-making, which will take into account "the ecological, cultural, social and economic situation in the entire Central Rockies Ecosystem."[33] This "big picture" or holistic approach is essential because Banff National Park is not large enough, for example, to sustain populations of large mammals such as grizzly bears and wolves.[34] Provincial lands beyond park boundaries, where bear hunting, logging, cattle grazing and resource extraction take place, are very different regimes from those of federal parks. Hence, ecosystem management demands more than the maintenance of parks as protected remnants. The challenge is to integrate management within and beyond park boundaries:

> The problem is that the park's boundaries are artificial. Plants and animals don't recognize lines drawn on a map. We've been slow to understand that setting aside a small postage stamp in the middle of a vast and complex ecosystem just doesn't work.[35]

---

[32] Parks Canada, *op. cit.*, 9.
[33] *Op. cit.*, 10.
[34] A male grizzly requires up to 2,000 km^2 to survive; while females need 200-500 km^2.
[35] Chief Warden Bob Hanley, as quoted in J. Krakauer, Rocky Times for Banff, *National Geographic* 188(1), 1995, 59-60.

Thus, while parks may serve as icons of environmental value, natural beauty, national pride, etc., they cannot fulfill conservationist expectations even within their own limited borders unless their connection to wider systems is recognized, and environmental issues outside parks are simultaneously addressed.

Closing Reflections

The new management plan is promising, but it will at best mitigate, not eliminate, the inherent incompatibilities between "natural" ecosystems and intensive tourism. Technical solutions and policy fixes can go only so far in restoring ecosystem components whose minimal spatial requirements have been undermined, and whose temporal rhythms have been truncated. Paradoxically, public interest in wilderness areas is a political necessity for their preservation; yet that same interest, translated to tourism on location, threatens to undermine the very value that inspires it.

The more we intervene in nature, it seems, the more urgently we seek to preserve some image of wilderness "untouched by human hand." Our impulse to preserve "intact wilderness," "ecosystem integrity," etc., seems a healthy reaction to the dangerous momentum of our technological interventions. Few ecosystems over the past several thousand years have been insulated from the effects of human action. Humans have always made consequential choices about our relations within ecosystems – with which components will we co-exist; which others will we transform or extinguish? To the extent that we value co-existence, and because we often cannot be sure of the effect of our choices, societal self-restraint is essential. Parks are an important locus for the exercise of such restraint, which must surely be generalized to wider environments.

Questions

1. If an ecosystem-based approach to park management requires an integrative strategy with respect to areas both within and beyond park boundaries, what obstacles and opportunities present themselves to achieving such an approach?

2. To what extent is it possible and or desirable to include ecological, cultural, social and economic values in the management of a national park?

3. What makes an ecosystem "natural," and what level of human activity is acceptable before it ceases to be "natural"?

Further Reading

Agree, J.K. and D.R. Johnson. 1988. *Ecosystem Management for Parks and Wilderness*. Seattle: University of Washington Press.

Dearden, P. and R. Rollins. 1993. *Parks and Protected Areas in Canada: Planning and Management*. Toronto: Oxford University Press.

Hummel, M., ed. 1989. *Endangered Spaces: The Future for Canada's Wilderness.* Toronto: Key Porter Books.

Lothian, W.F. 1987. *A Brief History of Canada's National Parks*. Ottawa: Parks Canada.

Woodley, S., Kay, J., and G. Francis. 1993. *Ecological Integrity and the Management of Ecosystems*. Ottawa: St. Lucie Press.

Web Sites

- Banff National Park: http://www.canadianrockies.net/poaching.html
- Banff and Jasper National Park: http://www.canadianrockies.net/JEM_Photography/photogr.html

- Banff National Park: http://www.worldweb.com/parkscanada-banff/zoning.html
- Friends of Banff National Park: http://www.worldweb.com/ParksCanada-Banff/friends.html
- Parks Canada: http://www.worldweb.com/parksCanada-banff/
- Parks Canada: http://www.worldweb.com/VertexCustomers/p/ParksCanada-Banff/books.html

Audio-Visual Material

- *All About Bears*, National Film Board, 1985 (dir.: Dennis Sawyer, 12 min.).
- *Great Days in the Rockies*, National Film Board, 1983 (dir.: Jerry Krepakevitch, Anne Wheeler and Tom Radford, 12 min.).
- *The Grizzly Bear: Losing Ground*, Canadian Broadcasting Corporation, 1991 (46 min.).
- *National Parks: Forever Wild?* Canadian Broadcasting Commission, 1994 (dir.: David Suzuki, 46 min.).

Case Nine
Canada Geese: Joint Management on the Mid-Atlantic Flyway

Focus Concept

International co-management efforts bring together federal, provincial and state governments, as well as researchers, indigenous resource users, and sport hunters in Canada and the United States, to cope with a sharp decline in the numbers of Interior Canada geese on the Atlantic Flyway.[1]

Introduction

Migrating geese have always made a dramatic impression on the human imagination, no species more so than the graceful Canada goose (*Branta canadensis*). Lines of geese aimed beyond distant horizons thrill the human soul, and evoke a longing for something inexpressible. Canada geese are many things to many people. For many sport hunters, they are the most prized of game birds. For many aboriginal people, especially in northern regions, they are a staple source of high-quality food, and symbolize a relationship to the wider environment that joins the material to the spiritual. For some urbanites and agriculturalists, coping with local instances of "over-population," Canada geese have come to represent a nuisance.

By their very nature as long-distance migrators, the conservation and management of Canada geese demands the cooperation of numerous interest groups, occupying several political jurisdictions. No one government, and no single human society, can function as sole stewards of such species. As a valued resource, Canada geese require management choices that are hotly contested by groups with divergent interests, particularly when the resource becomes less abundant. We can learn much about ourselves as a society, and how we manage our relationship to our natural surroundings, through analysis of such episodes.

Biological Attributes

The Interior race (*Branta canadensis interior*) is one of eleven recognized subspecies of Canada goose,[2] and predominates on a number of the central continental flyway populations, including the Atlantic flyway.[3] Canada geese on the nesting grounds in Ungava, northern Québec (Figure 9.1) are almost exclusively of the Atlantic population Interior subspecies, but a variety of geese from other populations and subspecies are also present on the broad migration corridor, and at wintering areas. These include Mississippi and Southern James Bay geese, Atlantic race geese (*Branta canadensis canadensis*), and geese of two small races (*Branta canadensis hutchinsii* and *Branta canadensis parvipes*). Giant Canada geese (*Branta canadensis maxima*) have a growing presence on the Atlantic Flyway. These geese are not

---

[1] This case study draws extensively on the research notes of Colin Scott, Dept. of Anthropology, McGill University, with whom this case study is co-authored.

[2] F.C. Bellrose, *Ducks, Geese and Swans of North America*, 2nd edition (Harrisburg, Pennsylvania: Stackpole Books, 1976); there is divergent opinion on the exact number of subspecies, with some authorities including many more in their classifications.

[3] Canadian Wildlife Service (http://www.ec.gc.ca/cws-scf/canbird/goose/goospop.htm), 1997.

Source: Adapted from U.S. Fish & Wildlife Service

Figure 9.1. Mid-Atlantic Flyway

arctic/subarctic breeders, but "resident" nesters in the St. Lawrence-Great Lakes region, where they have been introduced over the past three decades.[4] Flocks of Giant Canadas of pre-breeding age, however, do fly north to James and Hudson Bays in the spring where they remain for the summer molt, returning south in the fall. The dramatic success of the Giant Canadas, in contrast to the decline of Interior Canadas, has important management implications that will be discussed later in this case study.

Canada geese are renowned monogamists, typically forming life-long pair bonds around two or three years of age. Interior goose pairs nest in greatest concentrations on the Arctic tundra, but also in bog-fen habitat of the subarctic forest. Islets on ponds, which afford a good view of surroundings, are preferred nest sites, with clutch sizes of 4-6 eggs. While the female incubates, the male guards against territorial competitors and predators. Goslings leave the nest within a day of hatching and, accompanied by parents, progress from initial use of bogs and marshes to more open lakes, rivers and coastal areas. Fledging occurs after about two months, for the mid-sized races. Families remain together throughout one full migration cycle, flying together in the fall to wintering grounds on the mid-Atlantic coast, and returning together to the nesting grounds the following spring. The yearlings, not yet of breeding age, are evicted by their parents from the nesting area at the onset of breeding.[5]

The diet of the Canada goose is herbaceous, comprised of roots, bulbs, shoots and the seeds of grasses, sedges and other plants. In the coastal areas of James and Hudson Bays, the diet is dominated by grasses, sedges and other herbaceous saltmarsh plants . The seeds, roots, and bulbs of these plants are preferred in spring, whereas the shoots and leaves are more important in fall. Some freshwater plants, and berries of several species that are especially plentiful on tundra-like coastal islands and outlying peninsulas, are also eaten in fall.[6] Further south, waste grain left behind from farm harvesting has become an important food source for migrating and over-wintering geese.

Distribution and Migration

The migration of Interior Canada geese describes a wide corridor from the coastal and interior areas of the Ungava Peninsula, eastern Hudson Bay and James Bay, through southern Québec and eastern Ontario, to wintering grounds extending from New York to South Carolina on the Atlantic seaboard (Figure 9.1). Geese are hardy migrators; in the spring they reach southern James Bay while land and sea are still covered by snow and ice, with only a few ponds beginning to open up at creeks and rivers on foreshore flats, and at inland waterbodies. By June they have reached their nesting grounds, early enough that in difficult years late snow and ice cover, inclement weather and sharp drops in temperature can be important inhibitors of nesting success. In the fall, southward-migrating geese keep not far ahead of the advancing edge of winter, with some flocks remaining at staging areas along Hudson and James Bays until freeze-up forces them southward.

The pace of migration and the density of geese at specific points along the way vary dramatically from

---

[4] Resident geese include subspecies admixture, resulting from introductions of genetic stock from several parts of the continent through deliberate stocking and release of captive flocks; but the Giant Canada is genetically predominant.

[5] R.C. Cotter, P. Dupuis, J. Tardif, and A. Reed, Canada Goose, in J. Gauthier and Y. Aubry, eds., *The Breeding Birds of Québec: Atlas of the Breeding Birds of Southern Québec* (Montréal: Association Québecoise des Groupes d'Ornithologues/ Québec Society for the Protection of Birds/ Canadian Wildlife Society, 1996), 262-265.

[6] See A. Reed, R. Benoit, M. Julien and R. Lalumière, Goose use of the coastal habitats of northeastern James Bay, Occasional Paper Number 92 (Ottawa, Canadian Wildlife Service, 1996) for a detailed discussion of salt marsh, heath and eelgrass meadow habitat use.

season to season with conditions of weather, local food availability, etc.:

> Canada geese are very strong fliers and may travel more than 1,400 miles between wintering and breeding areas, and as much as 650 miles without stopping. Average migration speed is estimated at about 40 miles per hour but geese have been clocked at more than 90 mph when aided by tailwinds. Geese migrate at altitudes ranging from a few hundred feet to over 8,000 feet, depending on visibility and the location of the most favorable winds.[7]

Given these capabilities, it is possible for the main mass of a migration to completely overfly customary staging areas, when local conditions for feeding and resting are unfavorable, or when conditions further on are preferred.

Conservation Status of the Atlantic Population

Canada geese have adapted extremely well, on the whole, to the settlement and development of the North American continent by Europeans. The conversion of landscape to extensive regions cultivated for cereal grain as well as grassed pastures and parks, regulated hunting, and the creation of special sanctuaries and preserves, have resulted in a continental Canada goose population that may be greater than it was when Europeans first arrived.

The Atlantic Flyway population, however, is one of the few in North America that in more recent years has experienced quite serious population decline. Surveys conducted along the eastern seaboard of the United States indicated a population increase from 200,000 to over 900,000 geese between 1948 and 1981. Since the mid-1980s, however, the flocks are considered to be in serious decline, and by 1992-93 the surveys estimated around 570,000 birds.[8] This decline was considered all the more alarming because an increasing population of Resident geese included in survey counts masked what must have been an even steeper decline in the native Interior race.[9] Poor spring conditions on the breeding grounds, together with increased sport hunting kills, and possible competition from booming populations of Giant Canada and Snow geese, are considered the primary causes of the decline.[10]

Interior Canadas are genetically separate from Giant Canadas and other subspecies, and they have

---

[7] T.J. Moser, S.R. Craven and B.K. Miller, Canada Geese in the Mississippi flyway: A guide for goose hunters and goose watchers (United States Fish & Wildlife Service/Ducks Unlimited, n.d.); citing D.G. Raveling and H.G. Lumsden, *Nesting Ecology in the Hudson Bay Lowlands of Ontario: Evolution and Population Regulation*, Research Report 98 (United States Fish & Wildlife, 1977), and M. Owen, *Wild Geese of the World: Their Life History and Ecology* (London: B.T. Batsford Ltd., 1980).

[8] R.C. Cotter, *et al.*, *op. cit.*, 265. These numbers are under-estimates in as much as they include "observed" birds and are not extrapolated to provide estimates for wintering areas not included in the survey; but year over year, they are considered proportionally accurate indicators of population trends.

[9] Giant Canada geese have recovered from near extinction in the 1960s to what many consider a condition of over-population today. This subspecies has had remarkable success adapting to urban recreational and agricultural areas across a vast belt of the United States and southern Canada. Although migratory, they are migratory over relatively short distances. Numbers have increased to the point that public complaints about crop damage, excrement on golf courses and urban parks, and airport interference are rife. There are concerns that Giant Canadas are competing for scarce resources with other subspecies, such as migrant Interior Canadas, whose populations have been showing signs of decline on the Atlantic flyway and elsewhere.

[10] The Southern James Bay Population of Interior Canada geese has also suffered serious declines, and there may also be "early indications" of their decline in the Eastern Prairie and Mississippi Valley populations, according to the Canadian Wildlife Service (http://www.ec.gc.ca/cws-scf/canbird/goose/goospop.htm, 1997).

evolved a distinct adaptation to a unique combination of nesting and overwintering habitats. Their survival as a subspecies is an important contribution to the versatility and long-term viability of the species as a whole. For such reasons, efforts to protect biodiversity must take subspecies as well as species diversity into account, and it is important to manage flyway and subspecies populations individually.

Management responses to the Atlantic population decline, which are the focus of this case study, resulted in a significant recovery in the number of breeding pairs in 1996 (up 60 percent in comparison to 1995), and further increases in 1997.[11] Nesting success in 1996 was good despite the disadvantage of a late spring,[12] and early indications are that there has been excellent nesting success in 1997.[13]

The Migratory Birds Convention

To understand management responses to the specific problems of the Atlantic population, it is important to analyze the evolving legal and regulatory context of those responses. International treaty-making, federal-provincial-state jurisdictional arrangements, aboriginal claims settlements, and protection of Native rights in the Canadian Constitution and in Supreme Court decisions, are the key features of this context.

The Migratory Birds Convention (MBC) and subsequent attempts at amendment (centrally related to the constitutional requirement in Canada to respect Native hunting rights) provide an overarching regulatory framework. Signed in 1916, the Convention declares the responsibility of the Canadian and United States federal governments to manage migratory birds, with the goal "...of saving from indiscriminate slaughter and of insuring the preservation of such migratory birds as are either useful to man or are harmless..."[14] The MBC imposed a closed season on the hunting of nearly all migratory game birds (including geese) between March 1 and September 1. Furthermore, the open hunting season was not to exceed three and one-half months.[15]

There was a clear recognition of the specific rights and needs of aboriginal people at the time the Convention was negotiated,[16] and this is reflected in certain exceptions on their behalf: "...Indians may take at any time scoters for food but not for sale..." and "...Eskimos and Indians may take at any season auks, auklets, guillemots, murres and puffins, and their eggs for food and their skins for clothing, but the birds and eggs so taken shall not be sold or offered for sale."[17] The absence of exceptions for geese and ducks, among other species, would prove to be a major problem. It was recognized even in 1916, when Canada was writing its ratification legislation for the Convention (the *Migratory Birds Convention Act*), that changes would be needed to accommodate aboriginal rights and interests. Although aboriginal people made numerous requests as early as the 1930s that these changes be brought about, it was not until the 1970s that serious initiatives were undertaken. The issue was revisited during the negotiation with northern Québec Cree and Inuit of the *James Bay and Northern Québec Agreement* (JBNQA; 1975),

[11] W.F. Harvey and A. Bourget, A breeding pair survey of Canada Geese in Northern Québec (Maryland Dept. of Natural Resources/Canadian Wildlife Service), 1994-96.

[12] A. Reed and R.J. Hughes, Reproductive success and recruitment of Atlantic population Canada Geese in Northern Québec: A progress report for 1996 (Atlantic Flyway Council Technical Committee, 1997).

[13] Austin Reed, CWS, personal communication.

[14] *Migratory Birds Convention*, 1916, preamble.

[15] *Op. cit.*, Article II.

[16] Kathy Dickson, CWS, personal communication.

[17] *Migratory Birds Convention , op. cit.*, Article II.

the first comprehensive settlement of aboriginal claims of the modern era in Canada.[18] Under the terms of the JBNQA, Canada undertook to make best efforts to achieve amendments to the MBC.

The MBC Amendment Process

In 1979, a protocol to amend the Migratory Birds Convention was negotiated between Canadian and United States federal wildlife agencies – the Canadian Wildlife Service (CWS) and the United States Fish & Wildlife Service (USFWS). The protocol would have recognized a general exception for northern Native people to harvest migratory birds without seasonal and bag limit restrictions, subject only to extraordinary measures that might be needed in the event of a conservation crisis. Although the protocol was accepted by the United States Department of Interior, various interests (including sport hunting groups and some wildlife managers) were unconvinced of its merits. Ratification of the protocol in Senate was not promoted by the State Department, and the process languished.

In the meantime, circumstances in Canada were shifting. The *Constitution Act* (1982), the supreme law of the land, entrenched the aboriginal and treaty rights of the Indian, Inuit and Métis peoples of Canada.[19] Several more comprehensive claims settlements were signed during the 1980s and into the 1990s, each recognizing strong harvesting rights for subsistence purposes, and providing for direct involvement of Native representatives in wildlife policy and decision-making. Supreme Court of Canada decisions were tending to the conclusion that aboriginal people, whether or not covered by treaty or comprehensive claims agreement, could not be inhibited in their harvests of wildlife for subsistence purposes, provided that conservation was respected. Moreover, this aboriginal right takes precedence over recreational hunting and fishing.

The 1979 protocol, which referred in its amendments only to Native people of the north, had not taken into account the Métis, non-status people, and status Indians in Canada as a whole.[20] Subsequently, there were efforts to broaden dialogue on both sides of the international boundary, and to involve aboriginal people and other interest groups directly in the negotiations, in order to achieve a more complete protocol. In 1990 the Canadian Arctic Resources Committee undertook broad consultation at several public meetings across the country. Ninety-five different organizations, of which fifty-seven were aboriginal, attended. The recommendations emerging from this process endorsed the idea that the subsistence harvesting rights of aboriginal people must be accommodated, and gained the support of upper management at the Canadian Wildlife Service.

To obtain the backing of aboriginal people on proposed amendments, the CWS conferred with an advisory group of more than two dozen status and non-status Indians, Métis and Inuit. The group was assembled in 1993 and communicated regularly over the next year and one-half. The CWS also

---

[18] Many aboriginal people in Canada enjoy historic Treaty rights under agreements concluded from the early 1800s to the early 1900s. Large areas of Canada, including most of Québec, British Columbia, the Atlantic Provinces, and the Territories, were not included in these treaties, or were covered by treaties which evidence suggests were fraudulently obtained. These are the areas which are subject to so-called "comprehensive claims" negotiations, which began with the JBNQA, and will continue into the foreseeable future.

[19] *Constitution Act*, 1982, Section 35, including the *Canadian Charter of Rights and Freedoms*, 1982, Section 25.

[20] So-called "status" Indians are those who are registered under Canada's *Indian Act*. Non-status Indians are those whose ancestors were never registered, or who were removed from the federal registry according to genealogical and other criteria. Métis are people of mixed Native and European descent who, like non-status Indians, did not fulfill Indian Act criteria for registered status (usually due to the presence of non-Native paternity somewhere in the family tree), and who in certain regions developed a group identity somewhat separate from both European and Native ancestors.

undertook bilateral discussions with regional and national aboriginal groups. At the same time, it worked with the United States Department of Interior, and the International Association for Fish and Wildlife Agencies (IAFWA, whose members include representatives of all State regulatory agencies) to chart a course for protocol approval in the United States.

Several interest groups, beyond the federal governments and aboriginal groups, had to be taken into account: state and provincial government wildlife agencies, sport hunting associations in both countries, and conservationist organizations. In the United States, support from state agencies and their biologists was deemed crucial. Although federal jurisdiction is paramount, coordination with state regulatory agencies is needed, for example, to implement coherent policy on hunting restrictions, and states supply a large part of the management-relevant research. They therefore exert considerable moral and practical influence on federal decision-making: In the words of Canada's chief negotiator, "it was really important to have IAFWA behind us, because they could stop ratification by just whispering to friends... getting them on-side was very important."[21] Both CWS and USFWS efforts were instrumental in achieving this objective.

Canadian senior wildlife managers perceived the policy culture of their American counterparts to be "top-down" in character, by comparison to the more participatory, joint management measures to which Canada had committed itself in its dealings with aboriginal people. Certain problems of practical perception had also to be worked through. In the United States context, where geese are present for longer periods of the annual calendar, the idea that no closed season should be imposed on Native hunters seemed threatening. Acceptance was easier when it was understood that there are relatively small windows of time when northern Native people can actually harvest the passing flocks. Anxiety that Natives in Canada might be taking excessive harvests were alleviated through the presentation of quantitative research results on actual harvest levels. The support of the IAFWA was successfully obtained.

Within Canada, provincial support for the federal initiative was equally important. Because the original Convention was signed by Great Britain in right of Canada, and because migratory waterfowl are a transboundary, international matter, the provinces support federal jurisdiction. International conventions are ratified by the provinces, however. Provinces have jurisdiction for the land base, and the great majority of conservation officers on-the-ground are employed by the provinces. The provinces had a representative on the protocol negotiating team, and were uniformly supportive of the CWS position.

Sport hunters, together with some scientific wildlife managers, were critical of spring hunting rights for aboriginal people, on the view that there would be more geese returning on the fall migration if geese about to nest were not hunted. Whether conducted in the fall or in the spring, however, hunting decreases the breeding population. On one hand, it can be argued from a bio-energetics perspective that it is more costly to kill females at the point in their annual cycle when they have accumulated body reserves for the reproductive process. On the other hand, this argument excludes human resource users from the calculation of ecosystem efficiencies, a socially costly premise when dealing with peoples whose economies are dependent on wildlife harvesting. The real issue is a political one – allocation between user groups. Although sport hunters take the large majority of total geese killed, their harvests per hunter are much lower than those of northern aboriginal hunters, and this can lead to resentment.

Differences in Constitutional tradition are associated with particular ideological resistance in the United States to the notion that aboriginals should have greater harvesting rights than other citizens, a view that

[21] Patricia Dwyer, CWS, leader of the negotiations for Canada, personal communication.

was also heard in Canada. Associations of sport hunters in both countries were among the most vocal on this point:

> The Ontario Federation of Anglers and Hunters, for instance, were saying "if the Natives get the spring hunt, we should get it too;" they were having difficulty with fact that a group of people could be treated differently from another. In Alaska, the Constitution disallows unequal treatment... hunting groups in Canada were saying we should be like that. We were saying to the Americans that we respected their Constitutional practice, and they would have to respect ours.[22]

The contest between equality of individual citizens' rights, versus special collective rights for aboriginals, is a persistent one. It could yet pose political obstacles to final ratification of amendments to the MBC, although the upper management levels of government in both countries are now "on-side".

Formal negotiation of the protocol between the two governments was conducted in April of 1995. The unprecedented step was taken of including three aboriginal members on the Canadian negotiating team. They included, respectively, a First Nations representative from the Grand Council of the Crees of Québec with a history of involvement on the James Bay and Northern Québec Hunting-Fishing-Trapping Coordinating Committee;[23] a Northwest Territories Métis with connections to both the Congress of Aboriginal Peoples (representing non-status Indians) and the Métis National Council; and the president of the Inuit Tapirisat of Canada. The aboriginal members were not mere "observers," but members in full standing, with full roles in the final wording of the protocol. According to a senior CWS representative, "their voices were listened for the most and heard best" of all parties at the table.[24]

The protocol to amend the 1916 convention was signed December 14, 1995. It reaffirms the commitment of both countries

> to the long-term conservation of shared species of migratory birds for their nutritional, social, cultural, spiritual, ecological, economic, and aesthetic values through a more comprehensive international framework that involves working together to cooperatively manage their populations, regulate their take, protect the lands and waters on which they depend, and share research and survey information.[25]

The two countries commit themselves to meet regularly to review the status of migratory bird populations, the status of habitats, and to undertake special protective measures and arrangements as needed. Included among some fifteen conservation principles, in accordance with which bird populations shall be managed, is the "use of aboriginal and indigenous knowledge, institutions and practices."[26]

The protocol specifically cites both "conformity with the aboriginal and treaty rights of the Aboriginal peoples of Canada," and "the customary and traditional taking of certain species of migratory birds and

---

[22] *Op. cit.*

[23] The Grand Council of the Crees of Québec is the regional self-governing voice of the James Bay Cree, who are parties to the JBNQA, and the most intensive subsistence harvesters of Atlantic Population Canada geese. The Hunting-Fishing-Trapping Coordinating Committee is the wildlife co-management body for northern Québec, established pursuant to the JBNQA.

[24] P. Dwyer, *op. cit.*

[25] *Protocol between the Government of Canada and the Government of the United States of America for the Protection of Migratory Birds in Canada and the United States*, 1995, preamble.

[26] *Op. cit.*, Article II.

their eggs for subsistence by indigenous inhabitants of Alaska" as rationales for amendment.[27] The aboriginal harvesting right in Canada is formulated as follows:

> Migratory birds and their eggs may be harvested throughout the year by Aboriginal peoples of Canada having aboriginal or treaty rights, and down and inedible by-products may be sold, but the birds and eggs so taken shall be offered for barter, exchange, trade or sale only within or between Aboriginal communities as provided for the relevant treaties, land claims agreements, self-government agreements, or co-management agreements made with Aboriginal peoples of Canada...[28]

There are also provisions for non-natives in northern Canada to harvest for food throughout the year, where aboriginal claims, self-government, and co-management agreements recognize the right of aboriginal parties to permit such activity; for the "indigenous inhabitants" of Alaska to harvest migratory birds and their eggs, subject to seasons and regulations that "shall be consistent with the customary and traditional uses by such indigenous inhabitants for their own nutritional and other essential needs;" and for indigenous inhabitants of Alaska to have an "effective and meaningful" role, by participating on management bodies, in the conservation of migratory birds.[29]

One of the aboriginal representatives on the Canadian negotiating team expresses general satisfaction with the outcome of the process in the following terms:

> We were able to participate internationally in the protection of aboriginal and treaty rights, and to ensure that their present and evolving nature would be accommodated and respected. We think that if the protocol is ratified and subsequently leads to the modification of the Migratory Birds Convention and the enabling legislation, then these agreements and legislation will conform with our rights in regard to management and harvesting of the birds. It is not just an issue of harvesting, it is an issue of our rights of governance in the management of the birds.
>
> The protocol is important in other areas, for example with respect to protection of habitat and environment. There are provisions enabling the governments to act internationally to protect habitat. The protocol also includes provisions for review of the Convention, which is important because otherwise it becomes very difficult to amend the Convention as circumstances evolve... We also got recognition in the protocol of the requirement to use indigenous knowledge along with scientific research in managing migratory bird populations.[30]

In short, the protocol accommodates constitutional and administrative principles and practices as they have evolved in both countries in recent decades, while strengthening cooperation in research and conservation initiatives. The protocol has been ratified by Canadian Parliament, but still awaits ratification in the U.S. Senate, whose advice and consent are required prior to Presidential signature.

---

[27] *Op. cit.*

[28] *Op. cit.*, Article II.4.a.i.

[29] *Op. cit.*, Article II.4.a.ii, b.i-ii.

[30] Philip Awashish, Grand Council of the Crees of Quebec, personal communication.

Responses to the Decline in Atlantic Population Canada Geese

The handling of problems with Atlantic Population Canada geese illustrates in practice the application of principles and approaches negotiated in the protocol. From the mid-1980s to the early 1990s, broad consensus developed among wildlife managers that the Atlantic Population was in substantial decline.[31] The customary technique of mid-winter surveys yielded uncertain results with respect to Interior Canadas, however, due to the inclusion of an unknown proportion of Resident geese.[32] Annual breeding pair surveys of the nesting grounds in northern Québec were begun in 1993, to obtain a more discrete measure of the status of the Interior Canada geese population, and to provide more direct information about nesting success.[33] Methods chosen allowed comparison with an earlier survey conducted in 1988.[34] Aerial transects are flown over regions representative of primary nesting habitats: inland tundra, coastal tundra, and the transition zone between tundra and boreal forest.[35] The 1994 survey confirmed a marked decline in breeding pairs since the 1988 survey, and the 1995 survey brought further bad news, with estimates of breeding pairs down to about one-quarter the levels of 1988 (see Figure 9.2).[36]

Canada goose hunting was closed throughout 1995 and 1996 in all states and provinces where they were hunted during their migration and wintering periods. The closure was advocated by CWS biologists, and by the Atlantic Flyway Council (AFC), originally established as a vehicle for co-management between states and the United States federal government, and to provide a forum for the co-ordination of state input.[37] Six eastern Canadian provinces also now belong to the AFC. Membership includes the state wildlife departments, their provincial counterparts in Canada, and the two federal governments "ex officio." From time to time the CWS and various indigenous governmental entities have funded James Bay Cree and northern Quebec Inuit representatives to attend AFC meetings, where they participate as observers. The role of the Council was central in bringing about a moratorium on sport hunting of geese on the Atlantic flyway:

> Atlantic Flyway Council are basically biologists and government managers talking to each other; they can advise but have no power. But their recommendations carry some weight.

---

[31] L.J. Hindman and F. Ferrigno, Atlantic flyway goose populations: Status and management, *Transactions of the North American Wildlife and Natural Resources Conference*, vol. 55, 1990, 293-311. For an excellent historical review of research and research needs relating to Atlantic Population decline, see L.J. Hindman, R.A. Malecki, and J.R. Serie, Status and management of Atlantic Population Canada Geese, *7th International Waterfowl Symposium* (Memphis, Tennessee, February 4-6, 1996), 109-116.

[32] See G.E. Menkens, Jr. and R.A. Malecki, Winter sightings of Canada Geese, *Branta canadensis*, banded in northern Québec and James Bay, in R.A. Malecki and J.R. Serie, eds., *Canada Geese of the Atlantic flyway: A collection of published papers* (Canada Goose Subcommittee Technical Section, Atlantic Flyway Council, 1990), 65-67. These authors concluded that winter observations were insufficient to delineate breeding goose populations, due to substantial overlap in their use of winter area.

[33] D. Bordage and N. Plante, A breeding ground survey of Canada Geese in northern Québec - 1993 (Canadian Wildlife Service, Québec Region, 1993); W.F. Harvey and A. Bourget, A breeding pair survey of Canada Geese in northern Québec (Maryland Dept. of Natural Resources/Canadian Wildlife Service, 1994-96).

[34] R.E. Malecki and R.A. Trost, A breeding ground survey of Atlantic flyway Canada geese in Northern Québec, *Canadian Field Naturalist*, vol. 104, no. 4, 1990, 575-578.

[35] The 1988 and 1993 surveys had also included a boreal forest transect area, but this was dropped in subsequent years because it had relatively low densities of nesting geese with little annual variation; W.F. Harvey and A. Bourget, *op. cit.*, 1995, 3-4.

[36] Although spring weather conditions were more favorable in 1995 than they had been in 1994, the depressed levels were apparently the result of poor nesting success in 1992; Canada geese generally breed at about three years of age.

[37] The Atlantic Flyway Council is one of four such councils, each addressing management issues on a major flyway.

They were the group that recommended the closure of the sport hunt. It was very hotly debated, and a lot hung in the balance, because everybody had to be on-side for it to work.[38]

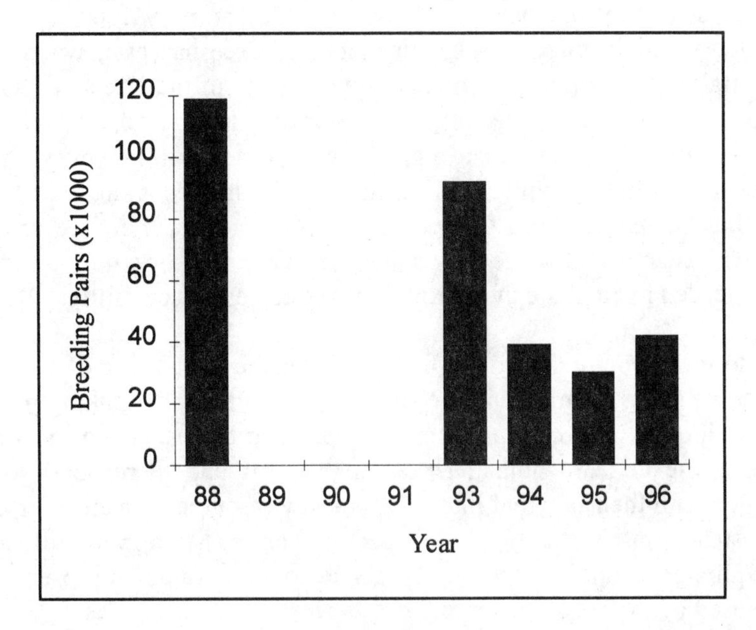

Figure 9.2. Index of the Number of Breeding Pairs of AP Canada Geese in the Ungava Peninsula of Northern Québec.

Source: Canadian Wildlife Service (Ottawa, http://www.ec.gc.ca/cws-scf/canbird/goose/cgat.htm, 1997).

At the same time, in 1996 regulatory agencies tried to implement hunts targeted specifically at over-abundant Resident Canadas. This is difficult, since sport hunters cannot readily distinguish Resident geese from migrants, but the approach is to allow hunting in the south after migrant geese have left for the north, and before they return in the fall.

Collaboration with Native Harvesters

It was widely understood by government managers in Canada that the impact of extending the moratorium to the Native subsistence hunt would be highly contentious, would impose hardship on Native communities, and could even be counterproductive for efforts to rebuild the Atlantic population.

Canada geese are exceptionally important, both economically and culturally, to the Cree and Inuit people of northern Québec. Geese are hunted for subsistence purposes during both the spring and fall migrations, mainly along the coastal corridor of James and Hudson Bays where the density of migrating flocks is the highest, but also at inland lakes and rivers. The harvest by Crees is the largest. Comprehensive recent harvesting data do not exist, but during five years in the 1970s the subsistence harvest by Crees in Québec averaged about 60,000 Canada geese annually, comprising approximately 16 percent of all harvested foodweight. In the communities along the coast of James and Hudson Bays, where dependence on waterfowl is heaviest, Canada geese accounted for 25 percent of all harvested foodweight.[39]

---

[38] Rick Cuciurean, Cree Trappers Association, personal communication.

[39] James Bay and Northern Québec Native Harvesting Research Committee, *The Wealth of the Land: Wildlife*

From the perspective of indigenous knowledge, geese present particular management requirements and limitations. The more sedentary populations of certain other primary food species, like moose and beaver, can be regulated mainly through self-adjustment of Cree harvests, where Cree hunting is the primary factor in mortality: "experienced hunters are well aware of the effects of human use of wildlife resources, as well as other variables affecting population trends, and adjust their decisions about the location, mix and intensity of harvests accordingly."[40] With waterfowl, however, Cree harvests are less significant than other factors in the population dynamics of the species, many of which occur in nesting, wintering, or migrating habitat far from Cree territory. Nonetheless, the Cree environmental ethic of "respect" for the birds demands that while they sojourn on Cree territory, they be able to feed, rest, and store up the body fat needed to continue their migration and to nest successfully:

> Cree hunters do make comprehensive efforts to foster the well-being of goose populations locally, by keeping stress from human predation to a minimum, by making provision for the refuge of geese through the rotation of hunting sites, and by respecting local nesting areas. The ability of geese to learn human predation patterns and to respond socially to that learning, coupled with their inherent mobility, is a key resource management issue from the Cree hunter's standpoint. Virtually every aspect of hunting strategy is influenced in some way by the requirement to minimize the geese's awareness of human presence, and to limit disturbance caused by the act of harvesting geese *per se*.[41]

An elaborate system of local knowledge and social practice underwrites these management goals, intended to maximize harvestability over the long-term by maintaining conditions in which the birds' needs are met. The symbolism and ritual practice surrounding geese, as with other harvested species, includes vivid reminders of the necessity for both efficiency and self-restraint in hunting.[42]

Estimates in the 1980s indicated that Québec Cree were taking 13.8 percent of flyway kills for the mid-Atlantic stock, and Québec Inuit 6.4 percent, while all other users, the overwhelming majority of them sport hunters, were taking 79.8 percent.[43] There was optimism, therefore, that adequate relief from hunting pressure could be obtained through the moratorium on sport hunting alone, if need be. The approach of scientific managers at the CWS was to

> cut off sport hunting in the south, where the bulk of the harvest occurs, and develop a multi-pronged research strategy in the north... We argued that if we imposed closure on subsistence hunting, it would also remove the possibility of gathering a lot of information.[44]

Harvests by the James Bay Cree, 1972-73 to 1978-79 (Québec City, JBNQNHRC, 1982), xi, 227, 230-232. For a summary and discussion of these findings, see C.H. Scott, cited *infra*.

[40] C.H. Scott, The socio-economic significance of waterfowl among Canada's aboriginal Cree: Native use and local management, in A.W. Diamond and F.L. Filion, eds., *The Value of Birds*, ICBP Technical Publication No. 6 (Cambridge, International Council for Bird Preservation, 1987), 49-62.

[41] *Op. cit.*, 55

[42] C.H. Scott, Science for the west, myth for the rest? The case of James Bay Cree knowledge construction, in L. Nader, ed., *Naked Science: Anthropological Inquiry into Boundaries, Power and Knowledge* (London: Routledge, 1996), 69-86.

[43] A. Reed, Harvest of waterfowl by the James Bay Cree in relation to the total kill of those stocks (Ste-Foy, Québec, unpublished paper, 1984); A. Reed and C. Drolet, Waterfowl harvest by Native people of northern Québec in relation to the total kill, in *Proceedings of the 1985 Northeast Fish and Wildlife Conference* (Hartford, 1985).

[44] Austin Reed, CWS, personal communication.

The CWS wanted to learn, for example, what proportion of the subsistence harvest was comprised of Interior Canadas, and what proportion was coming from other races that were not part of the declining population – in particular, the Giant Canadas. For this, there needed to be a subsistence hunt to sample, and Cree and Inuit participation in the research was essential.

Cree and Inuit had been asked by the federal Minister of Environment for their cooperation in curtailing their hunt in order to assist in rebuilding the population, and the request had been made public. Sport hunters in both the United States and Canada had been calling for a moratorium on Native harvests, to match that on sport hunting. There was some reaction from Native leadership in the press, emphasizing the economic and cultural importance of their hunt, and warning against infringement of their rights to harvest. At the same time, the Native organizations indicated their readiness to compare information on the status of the flocks, and to discuss voluntary measures to reduce their harvests.

In January of 1995, in a joint initiative that involved participation by leadership of the Grand Council of the Crees of Québec (GCCQ) and Cree Regional Authority (CRA),[45] the Cree Trappers Association (CTA), and CWS biologists, consultations were held in nine Cree communities.[46] With the assistance of maps, graphs and diagrams, the scientific evidence for negative trends in the migrant goose population was presented, and Cree expertise was solicited. Pursuant to these consultations, the CTA organized cooperation with the CWS on a morphometric study, as well as a program to improve the recovery rate of neck- and leg-bands from harvested geese, to help determine the actual proportions in the harvest of Atlantic Flyway Interior geese versus geese from other subspecies and flyways.[47]

The issue of reducing subsistence harvests was both practically and politically challenging. From the perspective of the individual, harvests per hunter had been in decline for several years – a circumstance probably due as much to increases in the population of hunters as to decline in the population of geese. It is considered unlikely by Native and non-Native observers that total harvest levels of geese have increased since the quantitative, region-wide surveys of the 1970s, despite a sharp increase in the population of Cree hunters; in other words, the average harvest per hunter has dropped steadily:

> According to a survey that we did independently in one community, we found out that even though there were three times as many hunters that year than before (the signature in 1975 of) the James Bay and Northern Quebec Agreement, the same (total) number of geese were killed... So it was difficult for us at first to accept the idea of cutting our goose harvest, but reluctantly... we had a general assembly in Waskaganish in one of the Cree communities, and we decided after two days of meetings that we would put a quota on ourselves.[48]

The Special General Assembly on Wildlife Management was held at the James Bay coastal community of Waskaganish in March, 1996, prior to the spring hunt. It consisted of Crees only, including

---

[45] The GCCQ and the CRA are, respectively, the political and administrative wings of Cree regional government.

[46] According to Kathy Dickson, CWS (personal communication), the view was that a local approach, involving direct communication with hunters, was needed; that communicating through formal joint management bodies was not in itself sufficient.

[47] Head, beak and tarsier measurements are sufficient for distinguishing Giant Canadas that fall above, or lesser Canadas that fall below, the normal size spectrum of Interior Canadas; but there is overlap in the normal size ranges of various subspecies. To further complicate the interpretation of measurements, Giant Canadas have interbred with other smaller subspecies. Therefore banding and band recovery is also crucial in ascertaining subspecies and flyway provenience.

[48] E. Gilpin, President, Cree Trappers Association, interview in *The Atlantic Population Canada Goose Research Program*, Canadian Wildlife Service, 1996 (VHS, 22 min. 44 sec.).

representatives from the local CTA and the Councils of each of the nine Cree communities. The matter was debated at length, and resulted in a resolution reflecting the complexity of the issue. The resolution in its preamble affirms Cree commitment to principles of both conservation and the protection of traditional hunting rights; cites terms of the JBNQA which shield Native hunters against "unilateral and unreasonably imposed regulations" by central governments, and which give priority to the maintenance of Native harvests over sport harvests when cuts are required for conservation purposes; reaffirms Cree responsibility for managing resources; and notes that "decisions of Canadian, United States and other governments and corporations continue to destroy wildlife habitat including waters, lands, forests... which decrease the habitat available to the animals, fish and birds."[49]

The resolution goes on to commit each Cree community to "adopt its own set of conservation measures" as part of "a regional plan which aims to reduce the individual and total harvest of Canada geese to one half of 1995 levels," to be overseen by traditional hunting group leaders. The resolution explicitly rejects the validity of any quota imposed by external governments as counter to "Cree traditional and constitutional hunting and trapping rights." Further, the resolution commits the CRA and CTA to establish a conservation and monitoring program, and undertakes to seek the cooperation of the Inuit in voluntary harvest reductions, including suspension of egg collection.[50]

No quantitative data exist to indicate whether the objective of a 50 percent reduction in harvests was in fact achieved. Cree and Inuit participation in the morphometric study and the bird-band retrieval initiative yielded important results, however. There was a significant proportion of Resident geese in the spring hunt (ranging from 10-40 percent at communities were reliable data were obtained), present at earlier stages of the migration, and extending further north, than had been anticipated.[51]

The multi-party community tour and the cooperation that flowed from it were judged a success by the CWS, who report an atmosphere of strong collaboration with the Cree and Inuit.[52] Early in 1997, a new federal-Cree consultation was held, with three objectives: a) to present the results of the measurements and band collection to hunters directly; b) to ask the Cree to maintain their commitment to a restrained harvest; and c) to seek renewed cooperation in morphometric research and band recovery.

Ongoing Research Program

With the cooperation of the Cree, the Inuit, and numerous governmental and non-governmental agencies in the United States and Canada, a multi-pronged research strategy was developed in 1996 to improve the information base for managing the Atlantic population Canada geese. The research projects include:

- surveys of the breeding grounds in northern Québec

[49] Special General Assembly on Wildlife Management, Canada Goose Resolution (Waskaganish, GCCQ/CRA, March 19-20, 1996); published in *The Nation*, vol. 3, no. 9, 1996, 20.

[50] *Op. cit.* The Inuit, through their Hunting, Fishing and Trapping Association, agreed with the request from the Cree, as reported in B. Batt, Native Canadians support Atlantic flyway goose restoration, *Flightlines*, vol. 3, no. 4, 1996, 7.

[51] The expectation had been that Resident Canadas were flying north later than the Interior migrants, toward the end of the migration, and that they would be concentrated in the more southerly portions of the James Bay/Hudson Bay coastline. Preliminary results are presented in R. J. Hughes, A. Reed, E. Gilpin, and R. Dion, Population affiliation of Canada Geese in the 1996 spring subsistence harvest in the James Bay region of northern Québec: A progress report for 1996 (Atlantic Flyway Council Technical Committee, 1997).

[52] Charles Drolet, CWS, personal communication; Austin Reed, CWS, personal communication.

- a study of basic breeding biology and annual production
- a morphometric study to determine subspecies affiliation
- satellite tracking of migration routes
- estimates of subsistence harvest, James Bay region (expanded in 1997 to Inuit communities)
- a comprehensive banding program, representative of the breeding range (started in 1997)

The program involves financial and in-kind support from the Canadian Wildlife Service, Arctic Goose Joint Venture,[53] Atlantic Flyway Council, Cree Regional Authority, Cree Trappers Association, Ducks Unlimited, Makivik Corporation, James Bay Energy Corporation, United States Fish & Wildlife Service, and the United States National Biological Service. The program proposal,[54] coordinated by biologists at the Canadian Wildlife Service,

> represents what we felt needed to be done in Canada. A lot of us felt that additional research should be done in the States regarding problems on the wintering grounds, and possible competition with other races of geese, but these ideas have not yet been followed up on. We set up a joint United States-Canadian committee to discuss what should be done in Canada, and assumed leadership in coordinating the research.[55]

The research, which will require approximately $500,000 annually to the end of the 1990s, had attracted funding in the order of $300,000 per year by mid-1997.[56] In an era of spending cutbacks by deficit-conscious governments, competing research priorities are a significant factor in the adequacy of science-based management. The maintenance of collaborative links between scientific researchers and local people in command of indigenous knowledge also requires significant financial support; both the CWS and the CTA have indicated that funding for this purpose is insecure.

Discussion and Conclusion

Recent research suggests that we can be "cautiously optimistic" about the recovery of Atlantic population Canada geese. The recovery is likely, in a few years time, to permit the restoration of sport hunting quotas in the south. If so, it is a management success story, one which was achieved within a framework of international treaty-making and respect for the hard-won constitutional and treaty rights of Native peoples. This respect increased the effective links between scientific and local knowledge inputs to management, and it seems likely that conservation efforts over the longer haul will benefit as a result.

Eloquent testimony to the value of this approach is found in the words in 1991 of a CWS biologist who played a key role in the government's response to declines in the Atlantic population:

> Each party can benefit enormously from the other's experience and knowledge. Wildlife biologists can bring a valuable quantitative dimension to the understanding of goose ecology and population dynamics, and transmit useful information on flyway-wide management concerns which are otherwise unavailable to the Cree. The Cree's intimate knowledge of the land and the geese can provide answers to many management questions or, at the very least,

[53] The Arctic Goose Joint Venture involves monies from the North America Waterfowl Management Plan, signed by the federal Canadian government in 1986.
[54] Arctic Goose Joint Venture Subcommittee, A program to address information needs for management of Atlantic population Canada geese, 2nd revision (Canadian Wildlife Service, 1996).
[55] Austin Reed, CWS, personal communication.
[56] Kathy Dickson and Austin Reed, CWS, personal communications.

> provide a sound background for planning and implementing scientific studies. With such an important base of common ground, and with the goose populations generally flourishing, the opportunities for developing cooperative management programs appear excellent.[57]

The growing strength of such views in the scientific community bodes well, if the means can be found to sustain an ongoing dialogue between scientific and local knowledge experts and managers.

Questions

1. What factors account for the relative success of joint management efforts to assist the recovery of Atlantic population Canada geese?

2. Should the collective rights of aboriginal groups take precedence over the rights of individual citizens to the recreational use of wildlife resources?

3. What does this case tell us about the optimum relationship between self-management and state-management of wildlife resources?

Further Reading

Canadian Wildlife Service/United States Fish and Wildlife Service, *Waterfowl Population Status*, 1996.

Cottier, R., P. Dupuis, J. Tardif and A. Reed. 1996. Canada Goose. In J. Gauthier & Y. Aubry, eds., *The Breeding Birds of Québec*. Montréal: Association Québecoise des Groupes d'Ornithologues/Canadian Wildlife Service, Environment Canada, Québec Region, 262-265.

Harvey, W.F. and A. Bourget. 1994-96. Breeding Pair Surveys of Canada Geese in Northern Québec.

Hindman, L.J., R.A. Malecki, and J.R. Serie, Status and management of Atlantic Population Canada Geese, *7th International Waterfowl Symposium*, Memphis, Tennessee, February 4-6, 1996, 109-116.

Malecki, R.A. and J.R. Serie, eds. 1990. *Canada Geese of the Atlantic Flyway: A Collection of Published Papers.* Contribution of the Canada Goose Subcommittee Technical Section, Atlantic Flyway Council.

Reed, A., R. Benoit, M. Julien and R. Lalumière. 1996. Goose Use of the Coastal Habitats of Northeastern James Bay. Occasional Paper Number 92. Ottawa: Environment Canada, Canadian Wildlife Service.

Scott, C.H. 1986. Science for the West, myth for the rest? The case of James Bay Cree knowledge construction. In L. Nader, ed., *Naked Science: Anthropological Inquiry into Boundaries, Power and Knowledge.* London: Routledge, 69-86.

Web Sites

- Canadian Wildlife Service: http://www.ec.gc.ca/cws-scf/canbird/goose/goose.htm
- Bird Monitoring: http://www.fws.gov/~r9mbmo/statsurv/mntrtbl.html#tbl
- Ducks Unlimited Canada: http://vm.ducks.ca/prov/DUCONT.HTM
- "Nuisance" geese: http://www.emagazine.com/4curgees.html
- Coalition to Prevent the Destruction of Canada Geese: http://www.icu.com/geese/links.html
- Listing of State wildlife management agencies: http://www.netexpress.net/~gideon/geese/agencies.html

Audio-Visual Material

- *The Atlantic Population Canada Goose Research Program*, Canadian Wildlife Service, 1996 (23 min.).

---

[57] A. Reed, Subsistence harvesting of waterfowl in northern Québec: Goose hunting and the James Bay Cree, *Transactions of the 56th North American Wildlife and Natural Resource Conference* (1991), 347-48.

Case Ten
Hamilton-Wentworth: A Sustainable Community Initiative

Focus Concept

A sustainable community initiative is found to have greater success in gaining participation at the regional government level than at the level of grassroots community and individual citizens.[1]

Introduction

Hamilton-Wentworth is a regional municipality located in Canada's manufacturing heartland. For many Canadians, mention of Hamilton, the municipality's industrial urban center, conjures up images of a city dominated by smokestacks, notorious for abuses of air and water quality. For those who regard the concept of a 'sustainable city' or 'sustainable urban community' as an oxymoron, Hamilton seems a prime example of the contradiction.

In recent years, however, Hamilton-Wentworth has not only shed this reputation, it is setting a standard for local sustainable development that is gaining recognition both nationally and internationally.[2] Since 1993, more than 300 communities and agencies from over 40 countries have requested information about Hamilton-Wentworth's sustainable community activities, while delegations from Argentina, Brazil, China, Gaza, Italy, Jamaica, Mexico, Russia, the Ukraine, Vietnam. Taiwan, the Philippines, and Thailand have visited to learn first hand about the Hamilton-Wentworth experience.

In October 1993, the Region of Hamilton-Wentworth was selected as the sole Canadian municipality and one of only fourteen communities around the world to serve as a Local Agenda 21 Model Community. In December 1994, the Region received from Environment Canada the Canadian Award for Environmental Achievement in the category of leadership by a municipal government.

Hamilton-Wentworth was also selected by Canada's National Round Table on the Environment and the Economy as an example of how to take steps towards sustainability at a local level. Members of the Round Table visited Hamilton-Wentworth in May 1994 to meet with people involved. International invitations have been received for Hamilton-Wentworth representatives to present their experience at conferences in Australia, England, France, Italy, Japan, the United States, the United Arab Emirates, and Turkey. Hamilton-Wentworth was chosen by the United Nations Commission on Sustainable Development as one of nineteen municipalities from around the world, to be featured as examples of best practices in sustainable community planning at their meetings in New York in April 1995.

Hamilton-Wentworth was also the venue for the environmental summit of the G-7 nations in May 1995 because of its progress in creating a sustainable community and implementing the goals of Agenda 21. Last year, it was one of the forty finalists from which twelve award-winning communities were selected by the United Nations Habitat II Secretariat. The forty finalists were profiled by the United Nations as role models for sustainable community activities at Habitat II in Turkey in June 1996.

---

[1] This chapter draws extensively on reports and documents which were kindly made available by the Regional Municipality of Hamilton-Wentworth. Discussions with staff involved in VISION 2020 were also very useful.

[2] Regional Municipality of Hamilton-Wentworth, Hamilton-Wentworth's VISION 2020 Sustainable Community Initiative, Project Overview Fall 1989 to Winter 1997, Prepared by The Strategic Planning Division, Regional Environment Department, Regional Municipality of Hamilton-Wentworth, March, 1997, 43-44.

Hamilton-Wentworth's success is mobilizing government, industry, and citizens to address a range of environmental, economic, and social concerns. Within the Regional Council, a number of projects have been initiated to implement its VISION 2020 Sustainable Community Initiative. These range from revisions of long-term planning and policy documents to specific initiatives such as the Bicycle Commuter Project. At the community level, there are encouraging signs that the level of awareness of sustainability is increasing, although a greater commitment at the individual citizen level is needed.

The Hamilton-Wentworth Region

The Regional Municipality of Hamilton-Wentworth, located at the western end of Lake Ontario, comprises six municipalities and includes the cities of Hamilton and Stoney Creek, the towns of Ancaster, Dundas and Flamborough, and the township of Glanbrook. It covers a geographic area of 111,300 hectares and is home to 468,440 people.[3] Approximately seventy-five percent of the Region's population are concentrated within the City of Hamilton, which represents just over ten percent of the Region's land area.

In the heart of Canada's industrial core, Hamilton's manufacturing, particularly of steel, has traditionally been the dominant force in the local economy. This situation is beginning to change. Today, the economy of Hamilton-Wentworth is increasingly fuelled by growth in the advanced manufacturing, advanced technology, environmental, food processing, health care, and small business/entrepreneurial sectors. Despite a reduction in employment in the local steel industry by almost 1,000 jobs a year on average for the last ten years, the Region has the lowest unemployment rate of major urban centres in Ontario. The recent privatization of the Hamilton International Airport and its new focus on cargo is anticipated to act as a further catalyst for growth.

The Region exhibits great diversity geographically, socially and economically. Almost 50 percent of the land area is prime agricultural land, and a further ten percent is designated as environmentally sensitive areas (ESAs). The Niagara Escarpment, designated by the United Nations as an internationally significant ecosystem, dominates the topography of the Region. Referred to as the "mountain" in Hamilton-Wentworth, this steep cliff cuts through the region at an average height of 91 m above Lake Ontario and encircles the verdant Dundas Valley as it winds around the head of Lake Ontario.

The Hamilton-Wentworth Region is located at the northern end of a small ecological zone known as "Carolinian Canada". This zone, which encompasses most of southwestern Ontario, has a summer climate similar to that of North and South Carolina. As a result, a wide variety of plant and animal species occur which are not found elsewhere in Canada. The Region is also a popular stop over or "staging" area for migrating birds in addition to the many waterfowl that make Hamilton Harbour and the shores of Lake Ontario their permanent home. Over 160 species of birds are known to nest in the area.

The viability of farming as an occupation has become a major issue of concern in recent years as farmers experience increasing pressure to sell their land for urban or rural estate development. Farming practices have also come under some criticism as a result of increasing groundwater and surface water contamination. Urban growth, the majority of which takes the form of low-density urban sprawl, is causing a variety of impacts on the local ecosystem. These include the development of prime agricultural lands, development, disruption and degradation of natural areas, and increased air and water pollution.

---

[3] Census of Canada, 1996

VISION 2020: A Sustainable Community Initiative

In 1993, the Regional Council adopted VISION 2020, a set of goals and actions intended to move Hamilton-Wentworth towards sustainability. The initiative is part of a comprehensive program arising from the principles and recommendations outlined in Agenda 21.[4] The community's involvement in identifying its vision of a sustainable future, the related implementation strategy, and a community-based monitoring program are a response to the objectives established in each of the four sections of Agenda 21. These include social and economic dimensions (Chapters 4, 7 and 8); conservation and management of resources for development (Chapter 10); strengthening the role of major groups (Chapters 27, 28 and 30), and means of implementation (Chapters 36 and 38).

Process

Hamilton-Wentworth's Sustainable Community Initiative started in 1989, when Regional Council decided that new mechanisms were needed to improve coordination between municipal policy goals, objectives, and budget decisions. At the same time, the Region's Official Plan and Economic Strategy were in need of comprehensive review and questions were raised about the principles that should guide revision. Based on research undertaken by the Planning and Development Department, "sustainable development" was identified as the appropriate guiding philosophy because of its emphasis on public participation and the achievement of a balance between economic, social and environmental concerns.[5]

In June 1990, the Regional Council formally launched the Sustainable Community Initiative by creating a citizen's Task Force on Sustainable Development. The group was instructed to develop, in cooperation with its fellow citizens, the concept of sustainable development as a basis for a community vision to guide future development in the Hamilton-Wentworth Region. Following the model of the successful Remedial Action Plan (RAP) for Hamilton Harbour established in 1986, the Task Force was set up as a multi-stakeholder roundtable comprised of eighteen individuals representing the following sectors: agriculture, business, community organizations, education, health services, labour, natural environment, social services, and urban development. Members of the Task Force agreed at the outset that all decisions and recommendations would be by consensus.

The Task Force was given two years to complete its mandate, which included the establishment of "a public outreach program to increase awareness of the concept of sustainable development and to act as a vehicle for feedback on potential goals, objectives and policies for the Region."[6] Over the next two years the Task Force met with over 1,000 citizens. A variety of methods were used to inform and involve the community, including town hall meetings, community forums, workshops, focus groups, and various promotional and advertising activities. The latter included the use of local print, radio and television media, information displays at local shopping malls and other locations, newsletters distributed to every household, direct mailings and a special telephone "ideas line".

Once the issues of concern had been identified, the Task Force turned its attention to the development of a community "vision" called *VISION 2020: The Sustainable Region.* This describes in broad terms, the type of community that Hamilton-Wentworth could be in the year 2020, if the community's actions follow the principles of sustainable development.

The goals and major directions identified during the extensive public outreach program were synthesized to form a broad implementation strategy that is presented in *Directions for Creating a Sustainable Region*

[4] United Nations Conference on Environment and Development (UNCED), Rio de Janeiro, 1992.

[5] Regional Municipality of Hamilton-Wentworth, *op. cit.*, 17.

[6] *Op. cit.*

and its companion document entitled *Detailed Strategies and Actions for Creating a Sustainable Region*. These documents attempt to identify the major policy shifts and decisions required by Government, community groups, businesses, and individuals to make VISION 2020 a reality. The reports include more than four hundred recommendations and specific actions that were identified as possible approaches to bringing about these shifts. Eleven key areas of policy change were highlighted. These are summarized in Table 10.1, with examples of some of the four hundred action recommendations. VISION 2020 was unanimously adopted by Regional Council on February 2, 1993, as a guide to all its decision-making.

Table 10.1. Goals and Specific Actions of VISION 2020

VISION 2020: Goals	**Examples of Specific Actions:**
Protecting a system of interconnected natural areas:	• in cooperation with all stakeholders, identify the hierarchy of natural areas and corridors, together with policies controlling land uses within and around the system of natural areas
Improving the quality of water resources:	• implement a user pay concept by metering all water users
Improving air quality:	• develop a minimum standard for the amount of vegetation required on residential lots
Reducing waste:	• establish a waste exchange depot
Reducing energy consumption:	• continue the conversion of public vehicles to alternative low polluting fuels.
Creating a more compact and diverse urban form:	• designate a firm urban boundary as part of the Official Plan, beyond which urban development will not be permitted
Changing our mode of transportation:	• consider the purchase of buses that can accommodate wheelchairs
Ensuring good health and well being:	• undertake efforts to ensure an accessible, affordable, nutritious and personally acceptable supply of food, safe drinking water and housing for everyone.
Enhancing the local economy:	• create a centralized resource centre to support people wishing to start a business
Empowering the community:	• hold regular town hall meetings and other forums to facilitate citizen input
Supporting the local agricultural sector:	• permit the direct sale of farm produce to the public

Source: adapted from ICLEI, Case study of the Regional Municipality of Hamilton-Wentworth (Local Environmental Initiatives [Toronto] Inc. and UN Commission on Sustainable Development, February 1995), 6.

Community-based Indicators

In the summer of 1994, Hamilton-Wentworth began an indicators project called *Signposts on the Trail to VISION 2020*. The goals of this project were threefold: to develop a system of indicators and targets that would assist and improve the way decisions are made; to increase awareness and understanding of VISION 2020 and sustainable development; and to increase the accountability of decision makers and the community.

Rather than relying on technically-based measurements for determining performance, the Region went back to the community. Using a number of focus groups and information centres to solicit input from the community, a set of indicators was developed. Five criteria were used to select possible indicators: measurability; ease and cost of data collection; credibility and validity; balance; and potential for affecting change. The objective of this approach was to avoid top-down assessment and to provide the average citizen with measures of progress that can be readily understood and actively responded to:

> Hamilton-Wentworth has developed a singularly innovative approach for monitoring progress that moves beyond merely reporting on the success of past activities; rather, it also forms the basis for inspiring and motivating further change, and for providing citizens, political leaders, business and community groups with the information they need to make sustainable decisions.[7]

A set of twenty-nine indicators, linked to the eleven theme areas of VISION 2020, are presented on Report Cards to the community at the Annual VISION 2020 Sustainable Community Day (Table 10.2). The first annual Sustainable Community Day was held in June 1994. The event was attended by over 1200 citizens, and involved the participation of 120 local agencies and businesses, and over 175 volunteers.

Table 10.2. Community-based Indicators

Theme Area:	**Indicator:**
Natural Areas and Corridors:	• % significant areas protected • Total length of hiking trails
Water Resources:	• Suspended solids discharged into Hamilton Harbor • Water consumption – all uses (metered accounts) • Number of "All Beaches Open for Swimming" days • Amount of road salt used on regional roads
Air Quality:	• Number of good or very good air quality days • Number of complaint about air quality per year
Waste:	• Space used at landfill sites annually • Annual users of hazardous waste depot
Energy:	• Per capita residential electricity consumption
Urban Form:	• Office vacancy rate in downtown Hamilton • % listed heritage sites in region designated
Transportation:	• Annual transit ridership per capita • Total length of bicycle routes
Health and Well Being:	• % regional population receiving general welfare assistance • Low birth weight babies as percentage of total births • Annual applications for affordable housing • Hospitalization rate for falls by persons 65+ years • Library items borrowed by Juveniles • Crime rates
The Local Economy:	• Rate of meaningful employment • % labor force with post-secondary education
Community Empowerment:	• Voter turnout for municipal elections • Applicants referred by the volunteer center
Agricultural:	• Annual acreage in field crops • Annual approvals for rezoning from agriculture to urban land uses • Number participating in Environmental Farm Program

Source: Regional Municipality of Hamilton-Wentworth, Hamilton-Wentworth's Sustainability Indicators, Hamilton Wentworth's VISION 2020 Sustainable Community Initiative, Environment Department, 1996.

---

[7] ICLEI, Case study of the Regional Municipality of Hamilton-Wentworth (Local Environmental Initiatives [Toronto] Inc. and UN Commission on Sustainable Development, February 1995), 4.

Implementation

Regional Council has adopted VISION 2020 as a guide to all future decision-making in the Region. As a result of this commitment a number of actions have occurred to bring the decision-making process more in line with the community vision for Hamilton-Wentworth.

Following adoption of VISION 2020 in February 1993, Regional Council established the Staff Working Group on Sustainable Development to identify mechanisms for integrating the principles of sustainable development and the vision statement into the operations of the municipality. The group developed a *Sustainable Community Decision-Making Guide*, as a tool to assist Regional Council staff in the evaluation of all proposed and existing policies, programs, and policies. Council reports are now required to recognize the equal importance of "the three legs of the sustainable community stool: the economy, the environment, and social/health factors." A section called 'Sustainable Community Implications' is included in all such reports. Since June 1996, the Staff Working Group has been revising decision-making procedures to incorporate sustainability criteria into several of its activities, including grant applications, tendering and purchasing policies, and internal auditing procedures.

Regional Council has also integrated the goals of VISION 2020 into its decision-making by revising long range planning and policy documents. In June 1994, Regional Council adopted a new Official Plan, called *Towards a Sustainable Region*, which directly incorporates almost one hundred of the four hundred detailed recommendations made in the VISION 2020 reports. In November 1994, Regional Council adopted the *Renaissance Report*, a strategic plan for long-term economic development. The primary goal of the strategy is to incorporate sustainable development into community economic development activities. The *Regional Transportation Review* and the *Comprehensive Pollution Prevention and Management Plan* have also been revised to reflect the goals of VISION 2020.

In November 1996, the Regional Chairman and representatives of a majority of municipalities of the Hamilton-Wentworth Region, representing over eighty-five percent of the population, agreed on principles which will form the basis of a new system of local government in the area.[8] These include the creation of a single unified municipality or city of nearly one-half million people. All budgetary and administrative processes will be controlled by the city with the objective of eliminating duplication and overlap within the civic service. The legislation creating the new city will also specify that this new local government system will cost at least $30 m less than the present system, with a forty-four percent reduction in the number of municipal councillors, and a significant reduction in the tax burden for the vast majority of residents in Hamilton-Wentworth.

A number of the specific projects and programs, recommended by the Task Force, have also been implemented by Regional Council. These include the development of a bicycle commuter network and the construction of combined sewer overflow reservoirs that have improved harbor water quality to the extent that swimming may soon be possible for the first time in forty years.

In addition to these projects, Regional Council has focused on building community awareness. Programs, such as the *Children's Sustainability Fair* and the *Young Citizens for a Sustainable Future* have been developed in partnership with community organizations and are geared toward the youth of Hamilton-Wentworth. Efforts are made to have exhibits at major community festivals and staff are available to make presentations to interested community groups. Other activities include special events, such as the *Crazy Commute* when everyone is encouraged to leave their motor vehicle at home for the day.

---

[8] The Regional Municipality of Hamilton-Wentworth, Final Memorandum of Negotiations (http://www.hamilton-went.on.ca/memo01.htm, Nov. 29, 1996).

In early 1997, Regional Council approved the formation of the VISION 2020 Progress Team with the mandate to: review the progress the community has made towards the achievement of VISION 2020; consider ideas and suggestions about ways to move the community closer to becoming the sustainable community set out in the VISION; and recommend renewed directions and actions to Regional Council. Applications from citizens interested in serving as members of the Progress Team were being invited at the time of going to press.

The level of commitment by Regional Council to the adoption of the principles of sustainable development and the goals of VISION 2020 is being recognized at provincial, national, and international levels. Hamilton-Wentworth is the recipient of numerous awards and acknowledgements, and has become a role model for other municipalities seeking to adopt a more sustainable direction.

Community Response

Hamilton-Wentworth's VISION 2020 has been highly commended for "its comprehensive community consultation and consensus building approach."[9] More than 1000 citizens were directly involved in the public outreach program conducted over two and one-half years by the Task Force. The annual VISION 2020 Sustainable Community Day involves over 150 local community groups and businesses and attracted almost 4,000 citizens to a variety of events in 1996.

An evaluation of the extent to which the VISION 2020 project has reached the citizens of the community was conducted in early 1996 by the Environmental Health Program at McMaster University. The evaluation involved a telephone survey of a randomly selected sample of 250 households. Participants were asked a number of questions designed to determine their general awareness of VISION 2020 and the concept of sustainability. Almost twenty-five percent of the respondents claimed to have at least heard about VISION 2020 or sustainable development. This was a significant improvement from 1993 when an independent study by the Social Planning and Research Council suggested that only about five percent of the population had an understanding of the concept of sustainability.

Despite these successes and the overall achievements of the project, the level of community support and participation in VISION 2020 has fallen short of expectations. In particular, the numerous opportunities for community involvement and feedback through the public outreach program were often poorly availed of. For example, only 160 people attended the brainstorming sessions of the town hall meetings, while less than seventy people attended the Community Meeting where the vision statement was released for discussion. In addition, many of the organizations approached to organize a focus group did not want to participate, with the result that only eighteen of the fifty groups targeted contributed to the project.

A further example of this lack of community-level commitment arose during efforts to involve the community in the implementation process through a citizen's organization called, "Citizens for a Sustainable Community". Although the organization was set up and still exists today, "it has never been able to attract the attention of the community and expand beyond its original membership of about fifty members."[10]

Regional Council has expressed its concerns about the lack of community involvement:

> Unfortunately VISION 2020 is still seen as an initiative to guide the decision making of Regional Council as opposed to the decision making of everyone in the community.[11]

---

[9] ICLEI, *op. cit.*, 1.

[10] Regional Municipality of Hamilton-Wentworth, *op. cit.*, 29.

[11] *Op. cit.*, 10.

The extent to which this lack of grassroots support for the project can be attributed to the approach taken by Regional Council is difficult to measure. While efforts were made to ensure the success of the public outreach program, Regional Council believes that "more time should have been spent promoting the project and encouraging participation from the community."[12] In particular, the negative response from the local media and public to the VISION 2020 goal statements is seen as a failure of the communication strategy to adequately inform the community of the Task Force's mandate and approach to achieving its mandate. A community led review has been proposed for 1998 that will attempt to address the issue of creating stronger community ownership of the Vision. The possibility that this lack of community-level commitment to the project is part of a more fundamental problem is discussed below.

Discussion

Over the past ten years, there has been an emerging consensus over the main ingredients for a sustainable city. Table 10.3 outlines the approach recommended by Agenda 21. Grassroots involvement and participatory approaches are widely regarded as central preconditions for bringing about permanent beneficial change. In fact, community empowerment has become a catch phrase within sustainability initiatives. Despite this, it is only in recent years that community consultation has become an integral part of decision-making processes. For all the lip service paid to the ideal of community empowerment, progress remains patchy. This is in part because effective means of community participation are only gradually being developed. The need to find mechanisms to increase community involvement has been highlighted by Ronald Doering, former executive director of the National Round Table on Environment and Economy:

> Sustainability planning must be community-led and consensus-based because the central issue is will, not expertise... We can't protect ecosystems, let alone restore them, unless ways and means can be found to integrate the work of all the communities within the region... We must ... experiment with ways that involve citizens more directly and deliberately into policy making at all levels. Ultimately it comes down to social and political will.[13]

From preventing problems arising to remedying those that have already arisen, working from the bottom up is a vital component of any urban environmental strategy. Most governments readily acknowledge this. However, in the absence of clear guidelines about what community empowerment and increased public participation actually entails, the approaches taken have mixed results. It is also worth noting, nonetheless, that while community groups may have been slow to respond to the challenge of sustainable development, neighbourhood-level organizations for resisting undesirable developments have grown in strength in recent years. There are also a growing number of examples of well run, locally sensitive projects tapping into local knowledge and resources to improve local environmental conditions. These are most successful when the projects have clear objectives and tangible outcomes, such as improving eyesore land spots, removing graffiti and picking up litter.

This tendency to improve or maintain an existing local environment is often tied to the NIMBY ('Not In My Back Yard') syndrome. Too often the greatest level of community commitment is associated with an avoidance of responsibility for activities deemed necessary but undesirable, such as building hostels for the mentally ill and finding locations for waste disposal. All too often affluent neighborhoods that have

---

[12] *Op. cit.*, 21.

[13] R. Doering, Sustainable communities: progress, problems, and potential, National Round Table Review, Environment Canada Special Issue on Sustainable Communities, Spring, 1994.

been able to mobilize resistance to such intrusions are motivated more by desire to maintain property values than improve the general environment.

Table 10.3. Suggested Initiatives for Sustainable Urban Management (Agenda 21)

1. Institutionalize a participatory approach to sustainable development based on a continuous dialogue between all actors, especially women and indigenous peoples.
2. Promote social organization and environmental awareness through community participation in identifying and meeting collective needs such as infrastructure provision, enhanced public amenities, and restoring the fabric of the built environment.
3. Strengthen the capacity of local governing bodies to deal with environmental challenges, especially through comprehensive planning which recognizes the individual needs of cities; promote ecologically sound urban design practices
4. Participate in international 'sustainable city networks'.
5. Promote environmentally sound and culturally sensitive tourism.
6. Establish mechanisms to mobilize resources for local initiatives to improve environmental quality.
7. Empower community groups, non-governmental organizations and individuals to manage and enhance their immediate environment through participatory approaches.

The seeming consensus on the basics of managing sustainable development tends to belie the practice. Sustainability may simply be an ideal that is unattainable within today's implicit assumptions of how we live our lives. Changing the way people behave becomes a fundamental requirement of sustainable development. To do this, concepts of quality and value are needed which transcend today's emphasis on materialism, and that which is close-at-hand and visible. Macro-initiatives, like regulating urban growth to accommodate the natural environment, or engineering urban form to encourage reduced energy consumption, are important improvements. But changes are also required to the way cities are governed and to ways which people acknowledge and respond to their individual responsibilities for environmental stewardship. The Hamilton-Wentworth Region is already making some significant changes to its system of local government. It has also gone some considerable distance to increasing its citizens' awareness and understanding of sustainability. Town hall meetings, community forums, and annual community days play an important part in the process of change, but communication and education on the issue must achieve greater ubiquity. Fundamental changes to the attitudes and values system of each of us as individuals are needed. This reflects the conclusion of the British Columbia Round Table on the Environment and the Economy. They suggest that a central requirement for achieving sustainability is

> ... an acceptance of the fact that we, as part of society in general, have to change our attitudes about the environment. It will not be enough to be satisfied with a convenient blue box program. Everything we do will have to be done with the environment in mind. It will not always be convenient and it will often be a personal challenge.[14]

Questions

1. Hamilton-Wentworth's 'VISION 2020: The Sustainable Community' has adopted the three points of the sustainability triangle: environment, economy and society. Discuss the connections between these.

[14] British Columbia Round Table on the Environment and the Economy, *Sustainable Communities* (Victoria: British Columbia Round Table on the Environment and the Economy, 1991), 3.

2. In so far as 'VISION 2020: The Sustainable Community' focuses on the Hamilton-Wentworth Region, it applies a bio-regional approach to sustainable development. What are the merits and shortcomings of this approach?

3. When does a community become too large for a person to feel that their participation makes a difference? What other factors encourage or discourage the involvement of individuals in a community?

4. Compare the community-based management initiative outlined here with the various approaches taken at Temagami (Case Four). Discuss the similarities and differences.

Further Reading

British Columbia Round Table on the Environment and the Economy. 1991. *Victoria: Sustainable Cities.* Vancouver: British Columbia Round Table on the Environment and the Economy.

Campbell, S. 1996. Green cities, growing cities, just cities? Urban planning and the contradictions of sustainable development. *Journal of the American Planning Association* 62(3):296-312.

Doering, R.L. 1994. Sustainable communities: progress, problems and potential. *National Round Table Review* Spring, 1-3.

Hancock, T. 1996. Healthy, sustainable communities: concept, fledgling practices and implications for governance. *Alternatives* 22(2):18-23.

Maclaren, V. 1993. *Sustainable Urban Development in Canada: From Concept to Practice.* 3 Vols. Prepared for the Intergovernmental Committee on Urban and Regional Research. Toronto: Intergovernmental Committee on Urban and Regional Research Press.

Richardson, N.H. 1992. Canada. In R. Stren, R. White and J. Whitney, eds. *Sustainable Cities: Urbanization and the Environment in International Perspective.* Oxford: Westview Press.

Web Sites

- Business Coalition for Sustainable Cities: http://www.earthpledge.org/about/bcsc.html
- Sustainable Cities Program (SCP): http://unep.unep.no/unon/unchs/scp/general1.htm
- AHURI: Sustainable Cities (Australia): http://www.ahuri.edu.au/AHURI/research/resprior/SUSTCITY.html
- WHO Healthy Sustainable Cities Homepage (Europe): http://www.who.dk/tech/hcp/sustain.htm
- Urbanization and Sustainable Cities: http://biology.semo.edu/web/courses/bs105/urban.html
- Hamilton-Wentworth: http://www.hamilton-went.on.ca

Audio-Visual Material

- *The Niagara Escarpment: A Rock Video*, Canadian Broadcasting Corporation, 1986 (51 min.).
- *Wild in the City*, 1985 (16 min.).

Notes

✍ Notes

Notes

Notes

Notes

Notes

Notes

✍ Notes

✍ Notes

✍ Notes